MATERIALS SCIENCE AND TECHNOLOGIES

RECENT ADVANCES IN ABRASIVES RESEARCH

MATERIALS SCIENCE AND TECHNOLOGIES

Additional books in this series can be found on Nova's website under the Series tab.

Additional e-books in this series can be found on Nova's website under the e-book tab.

MATERIALS SCIENCE AND TECHNOLOGIES

RECENT ADVANCES IN ABRASIVES RESEARCH

DIRK BÄHRE
EDITOR

nova publishers
New York

NOTICE TO THE READER

Library of Congress Cataloging-in-Publication Data

Recent advances in abrasives research / editor, Dirk Bähre.
pages cm
Includes bibliographical references and index.
ISBN: 978-1-62808-566-2 (hardcover)
1. Abrasives--Research. Bähre, Dirk.
TJ1296.R43 2013
553.6'5--dc23
2013028247

Published by Nova Science Publishers, Inc. † New York

CONTENTS

PREFACE

Dear reader,

Under the term "Abrasives", we summarize particles which are characterized by a very strong hardness. Due to their size and form, they are suitable to produce material removal in contact with surfaces and have a high resistance to wear. Abrasives can be composed of different materials from the field of ceramic compounds, like e.g. oxides, carbides, and nitrides. The application of abrasives is very diverse. One focus is manufacturing technology, where abrasives are used as tool materials especially in technologies for the very precise machining of parts with smooth surfaces. Other fields of application are the removal of hard materials like rocks and the protection against wear, like e.g. in the case of brake discs.

The investigation of abrasives with their influences and effects in different stress situations is a very innovative, interdisciplinary field in science. All over the world, many research groups work on topics with the focus on abrasives. In this edited volume, different current results from the research in abrasives are compiled. They represent impressively the potentials abrasives have in technical applications. The variance of the methods and proceedings is as large as the fields of application. Different theoretical reflections and experimental analyses contribute to understanding the mechanisms and provide the basis for future-oriented developments. The results and research activities published in this book are intended to contribute to transfer findings worldwide and beyond the particular fields of application and to stimulate ideas for further research.

I am very glad that with this book Nova Science Publishers, Inc. offers a platform that represents the research in the field of abrasives in its entire scope. Thus, an international and interdisciplinary treatment of the topic is encouraged. I want to thank especially Ms. Carra Feagaiga and Ms. Jennifer Ramirez for the great collaboration in compiling the contributions of the book and for the always fast and uncomplicated resolving of all questions.

I thank all authors for their very innovative contributions and the research papers remarkable in every respect which are outlined here. Many thanks also to all those who participated in reviewing the contributions for their elaborate and constructive considerations. I especially thank Ms. Anne Gorius who managed all editorial tasks in an excellent way, participated in the linguistic revision and translated parts of this book into English. Without her great management this book would not exist.

With best regards,
Dirk Bähre

Saarbrücken, May 2013

Contact:
Prof. Dr.-Ing. Dirk Bähre
Saarland University
Institute of Production Engineering
Campus A4 2
D-66123 Saarbrücken, Germany
www.lft.uni-saarland.de

In: Recent Advances in Abrasives Research ISBN: 978-1-62808-566-2
Editor: Dirk Bähre

Chapter 1

QUARTZ-BASED COMPOSITES FOR USE IN ABRASIVE CROWNS OF ORNAMENTAL ROCKS

Marcello Filgueira*
Universidade Estadual do Norte Fluminense – UENF
Centro de Ciências e Tecnologia – CCT
Laboratório de Materiais Avançados – LAMAV, Califórnia
Campos dos Goytacazes/RJ, Brazil

ABSTRACT

Abrasive crowns are tools employed for the processing of ornamental rock plates, involving grinding, polishing and brightening. These steps have the abrasive grain size used at the crowns as the basic difference. The crowns are normally made of abrasive particles (diamond or silicon carbide – SiC) of varied sizes and immersed in a ceramic matrix (cement mortar), or a polymeric matrix (polyester or epoxy). The abrasive particles are responsible for the polishing action. Synthetic diamond powder is used when the matrix is polymeric, and silicon carbide is used when the matrix is SOREL (magnesian cement). Both ceramic and polymeric matrices perform like an agglomerating agent, retaining the abrasive particles inside the abrasive tool.

---

* Corresponding Author. Email: marcello@uenf.br

The present work develops the processing, characterization and abrasion wear tests of composites based on the system epoxy-silica, considering the silica - SiO_2 as quartz. The objective was to evaluate its technological application as abrasive crowns for the grinding of ornamental rocks, comparing them with those usually made in SOREL-SiC. Composites were produced with compositions epoxy – (70, 75, 80, 85 and 90 % weight of silica (mean grain size of 560 μm – equivalent to the first rock grinding step). The compounds were mixed and cured with a catalyst to produce the initial samples, and the crowns. The composites were submitted to abrasive wear tests and compressive tests, simulating the grinding operation of ornamental rocks. Surface finishing produced in an ornamental rock (grey granite) and the weight loss of the crowns were evaluated and compared to those obtained for commercial magnesia cement SOREL–SiC. Micro-structural analyses were performed accordingly. The work concludes for the positive technical viability of this new composite epoxy - quartz for use as grinding crowns in the ornamental rock industry, once these crowns were made fast and with simplicity. Among the processed composites, the composition epoxy-85 % quartz was considered to present the best results. Surface finishing, compression strength and wear resistance were as good as those from the commercial SOREL-20 % SiC crowns.

Keywords: Grinding, wear mechanisms, composites, quartz

1. INTRODUCTION

The study of materials for use as abrasive tools at the processing activities of ornamental rocks, including grinding, polishing and brightening had an expressive growth in the last decade [1], following the evolution experienced from the 80s up to now for rock cutting tools, like saw-discs, diamond wires, etc.

Abrasive crowns are tools employed for the processing of ornamental rocks plates, involving grinding, polishing and brightening. These steps have as basic difference the abrasive grain size used at the crowns. The crowns are normally made of abrasive particles of varied sizes and immersed in a ceramic matrix (cement mortar), or a polymeric matrix (polyester or epoxy). The abrasive particles are responsible for the polishing action. Synthetic diamond powder is used when the matrix is polymeric and silicon carbide when the matrix is SOREL – a magnesian cement. Both ceramic and polymeric matrices perform like an agglomerating agent, retaining the abrasive particles inside the polishing tool [2].

The tool begins to wear with the abrasive particle cambering (flattening or dulling of the abrasive grain). In the sequence, the matrix is also subjected to wear, reducing the area in which the abrasive particle can be retained. Increasing this wearing action up to a critical point, the abrasive particle is detached and a new layer of abrasive particles rises at the surface. This mechanism is reported in the literature as "pull-out" [3]. The oldest explanation for the wearing mechanism of an abrasive particle is that small fragments of the resultant surface are probably constituted by a series of asperities whose faces are plans of easy cleavage [4]. According with Wilks and Wilks [5], the wear rate did not increase with temperature, but is dependent on the abrasion direction – it is an important matter to be commented, because the contact friction between the abrasive particles and the stone generates a temperature as high as 1000ºC at the micrometric tip of the abrasive particles.

Nowadays, the wear mechanism that is more accepted for a cutting tool, based on diamond particles impregnated in an agglomerated matrix, was proposed by Wrigth and Wapler [6] and explained by Davis [7]. This mechanism is illustrated schematically in Figure 1, showing the behavior of the exposed dimension (protrusion) of a diamond particle drenched in the matrix (outside the matrix, responsible for the cut) plotted against time or area of cut stone. At region 1 is observed the initial matrix wear and the rising up of the abrasive particles cutting faces. Region 2 is called constant working zone, where a light abrasive wear of the agglomerated matrix is observed. Region 3 is characterized by the diamond dulling (flattening) followed by the breaking down of its cutting edges and the self-sharpening of the abrasive particle due to the continuous friction with the stone. Region 4 is characterized by the erosive wear of the matrix by the stone, with partial liberation of other abrasive particles' faces. Region 5, where the particle is broken, ends with the particle pull-out at region 6.

Regarding the unnumbered regions in Figure 1, which is only qualitative, some points must be commented: (1) in the left upper region, as much as region 1 approximates the protrusion axis, it indicates that the abrasive particles are weakly adhered by the matrix, what reduces the cut area and the cutting time. This is not verified in the practice because the protrusion height is controlled in such a way that a major part of the abrasive particles begin the cutting operation inside the matrix – approximately 2/3 of their height – it means that the mean protrusion height of abrasive particles in a new tool is by

1/3 of its size (mean diameter); (2) in relation to the central region, one must observe that the tool wear is a continuous process, so that this region is characterized by a tool with just one layer of abrasive particles. The tool in question possesses abrasive particles in the entire bulk; (3) related to the right upper region, one must consider that when the abrasive particle is flattened, its protrusion height is diminished – followed by a matrix wear, which in turn increases the particle protrusion height. Thereafter, the particle may suffer a direct and early pull-out (it also explains the absence of a frontier between areas 4 and 6), or continue the process until its breakage and posterior pull-out.

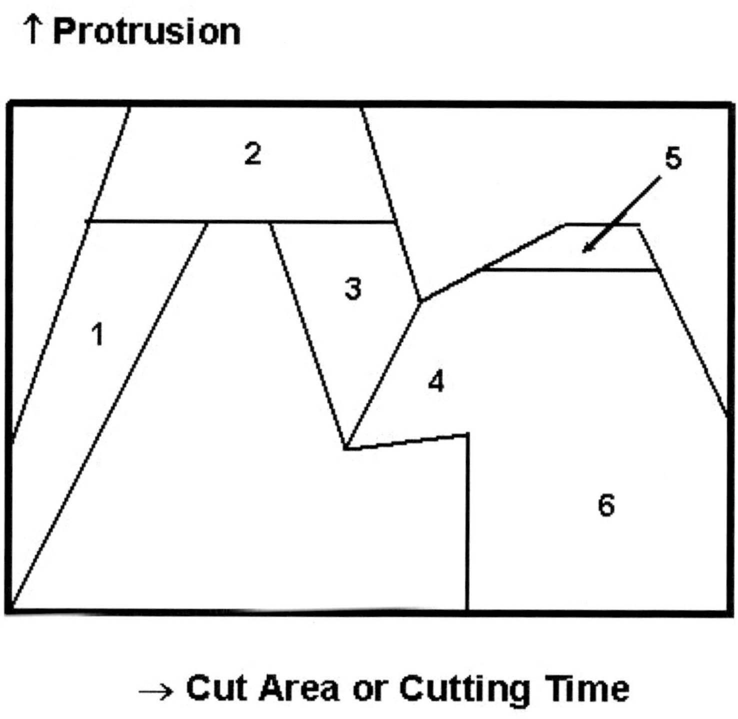

Figure 1. Hypothetic life time of a diamond particle (according toWright and Wappler [6]).

The use of abrasives into a solid surface results in material removal. From the point of view of the material to be worked, the wear mechanism depends on the mechanical properties of the solid. If the solid is ductile, the basic wear mechanism is related with plastic deformation. In addition, hardness is the parameter that governs the amount of material to be removed. However, the dominant abrasion mechanism for a brittle material depends predominantly on the fracture at the surface, in such a way that toughness is the dominant property [8].

When looking for the increase of abrasion resistance of a grinding, polishing and brightening tool based on composites, the addition of phases can be introduced to the system, within a ductile or brittle matrix. However, the mechanical properties required by the reinforcement are considerably different for both type of matrix. For a ductile matrix, one hard secondary phase is necessary in order to reduce the wear of the tool. On the other hand, when working with a brittle matrix, a tough reinforcement is necessary in order to increase the abrasion resistance because the presence of a tough phase reduces the composite tendency for surface fracture [8].

The objective of the present work is the processing and characterization of a new composite for the grinding operation of ornamental rocks. Its wear resistance as well as granite surface finishing quality was compared to available commercial crowns, and the wear mechanism was qualitatively explained. This new composite proposed is made using quartz powder (SiO_2), as the hard abrasive particles finely dispersed into a ductile matrix of epoxy. This new proposition for the material of the tool complies with the wear requirements for the grinding of ornamental rocks. The goal of this work is to offer one more contribution to this subject not too much exploited in the literature.

2. Methodology

2.1. Composites' Processing

2.1.1. Small Samples

Composites were processed using a liquid epoxy resin as the bonding matrix with abrasive particles of silica (SiO_2) finely dispersed into the matrix. The content of silica used was 70, 75, 80, 85 and 90 % weight.

The silica used is a beach sand, which was previously washed and screened. The used fraction was that retained between #20 and #40 mesh size sieves. Properties of the screened silica sand used are: rounded morphology (approximate aspect ratio = 2); mean grain size of 560μm – obtained using a RL New 21A sedigraph - considered adequate for the first step of ornamental rocks' grinding [2]; chemical analysis obtained in a Shimadzu EDX indicates the presence of 98.4 % SiO_2, 0.77 % Fe_2O_3, 0.39 % TiO_2, 0.34 % CaO, 0.23 % ZrO_2, 0.14 % P_2O_5, and other traces; and a density of 2.65 g/cm^3. This SiO_2 is in the structure of quartz.

The bonding agent was a liquid epoxy resin of the commercial brand 8003. This is a bisfenol A based resin, delivered by Epoxtec Produtos Quimicos Ltda, with the specific weight of 1.16 g/cm^3.

Twelve specimens of each composition of epoxy-quartz were manually prepared, with approximate dimensions of 15 mm in height and 12.7 mm in diameter. To accelerate the matrix cure, a liquid catalyzer offered by the same resin manufacturer was used, with commercial brand 21 Z, based on aliphatic amine and solvent free. For each 100 g of epoxy, 25 g of catalyzer were added, according to the manufacturer recommendation.

2.1.2. Grinding Crowns

After the complete characterization of the small samples, compared to samples from the commercial SOREL-SiC samples, it was found that crowns of the composites epoxy + (75-80-85) wt % quartz should be processed.

The grinding crowns fabrication process is described as follows:

1) obtention of industrial moulds by Cobral Limited, for the cure of the epoxy-quartz composites, with dimensions and geometries of the industrial grinding crowns;
2) sieving of the abrasive material – silica from beach sand: -20/+40 mesh = mean grain size of 560 μm – equivalent to the first rock grinding step;
3) mass fractions of silica (98 wt% quartz) and epoxy determinations for each composite: epoxy-(75-80-85) wt% quartz, as well as catalyzer (25 g for each 100 g epoxy);
4) homogenization into a cylindrical mixer by Pyramid Limited;
5) curing of the composites inside the industrial mould during 5 hours;
6) demoulding.

After the cure and demoulding, the processed crowns were submitted to industrial grinding tests in a stone processing industry.

2.2. Small Samples Characterizations

2.2.1. Wear Tests of the Small Samples

Seven small samples were submitted to wear testing, simulating the polishing operation of ornamental rocks. The rock used for the tests was an "Andorinha" gray granite, with oriented structure and coarse grain size (from 4 to 20 mm), petrographically classified as biotite monzogranite, containing 32 % quartz, 31 % microcline microperitic, 24 % plagioclase, 11 % biotite and 2 % accessories. Other properties of the gray granite are: a density of 2.67 g/cm^3, water absorption of 0.35 wt% and a total porosity of 0.93 vol%.

The ring of the gray granite had the following dimensions: outer and inner diameter, 80 and 40 cm respectively, and 5 cm high. The ring was horizontally setup on a table coupled to a micro processed physical simulator type AMSLER model AB800-E. Each specimen was vertically coupled to the granite ring throughout a support, delivering fixation and loading a weight over the small sample. The testing parameters were: vertical weight over the samples of 50 N, granite ring rotation of 40 rpm, running time of 2 minutes. During the experimentation, the granite ring turned at the required rotation, with the specimen remaining fixed at a point. Before and after testing, the granite ring roughness Ra was measured with a Mitutoyo rugosimeter with a 0.001 mm resolution. Measures were made at several points along all the perimeters of the abraded area, distant at least 5 cm one from the other.

Combining the sizes of the grains in the granite with the distance between measuring points, it is observed that the same grain may be assessed several times, depending on its orientation – it was intentional to ensure a higher reliability in the measurements, reflected by the small standard deviation observed in Table 1. Based on these results, Equation 1 was used to calculate the surface finishing (SF) of the granite disc and Equation 2 to calculate the weight loss (WL), this last corresponding to the abrasion resistance. See Figure 2 and 3 for a better understanding of the wear test procedure and the physical simulator.

$$SF = RB - RA \qquad (1)$$

where: RB and RA are, respectively, granite roughness (mm) before and after the abrasion test.

$$WL = WI - WF \tag{2}$$

where: WI and WF are, respectively, small sample weight (g) before and after the abrasion test.

Relative WL = SOREL 20 % SiC WL / small sample WL *(3)*

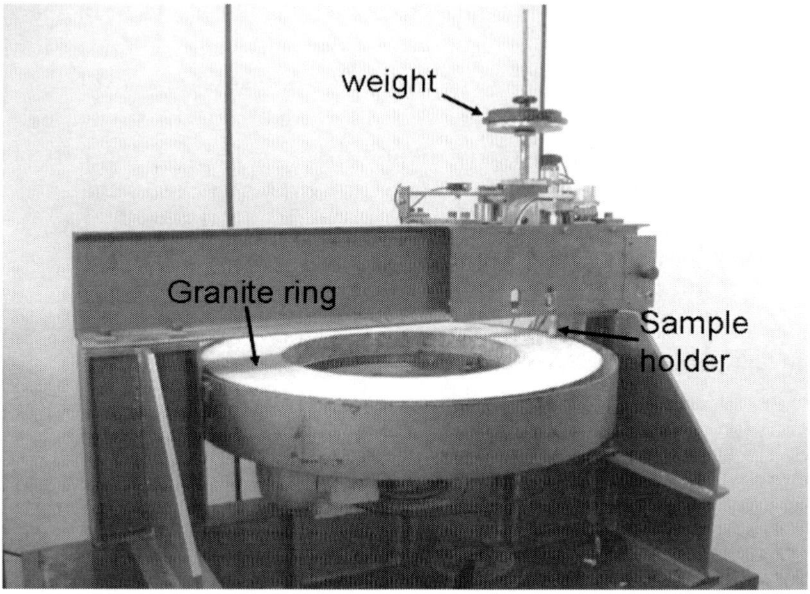

Figure 2. Physical simulator used for the SiO_2-epoxy and SiC-SOREL small samples wear tests.

2.2.2. Microstructure of the Small Samples

The microstructure of the surface of each composite was verified after the abrasion test with a scanning electronic microscope (ZEISS DSM 962), operated at 20 kV. Samples were gold sputtered coated to ensure good electrical conduction.

For comparative evaluation, specimens prepared from a commercial crown based on the magnesia cement (SOREL– 20% weight SiC) were also tested under identical test conditions as the epoxy-silica composites described above.

Figure 3. Wear test sequence conducted in the physical simulator for an epoxy-quartz small sample: (a) fixed sample at the beginning of the test; (b) and (c) continuous abrasion wear of the sample; (d) final of the test with the extensive wear of the sample.

2.2.3. Compression Tests

Five small samples were submitted to compression testing because during polishing the crowns are submitted to a mean load of 500 N. Compression tests were conducted using a universal tensile test machine (Instron 5582) with a speed of 2 mm/s.

2.3. Processed Crowns Characterizations

The industrial grinding tests of the processed crowns were performed at the Marmoraria São Salvador Limited – a stone processing company, located in Rio de Janeiro State/Brazil. These tests were performed using an "Andorinha" grey granite plate (1.5 m x 2.5 m x 0.03 m) which was divided in

4 equal parts – see Figure 4. Each part was grinded by one of the 3 processed crowns (group 1 – crowns number 1 to 6 mean 75 % quartz; group 2 – crowns number 7 to 12 mean 80 % quartz; group 3 – crowns number 13 to 18 mean 85 % quartz). The fourth group (crowns number C1 to C6) is regarded to the commercial SOREL-SiC crowns, for comparison. Figure 5 shows the grinding machine with the crowns on the rough surface of the granite plate.

Initially, dimensional measurements (height) of all 24 grinding crowns were performed, using a digital sliding caliper by Mitutoyo – 0.01 mm resolution, as well as mass measurements using a scaltec balance – 0.1 g resolution. After that, the industrial grinding tests were carried out, using a semi-automatic grinding stone machine. Each batch of crowns (group) was fixed in the machine satellite (6 crowns per test – see Figure 5). The tests duration was 3 minutes, based on the grinding machine operator experience. After this time, the dimensional and mass measurements were performed again, as well as during an intermediate time of 1.5 min. (90 sec.).

The roughness Ra of the granite plate surface was measured by a rugosimeter by Mitutoyo – 0.001mm resolution. Microstructural aspects of surface parts of the granite were observed in a Carl Zeiss Jena microscope.

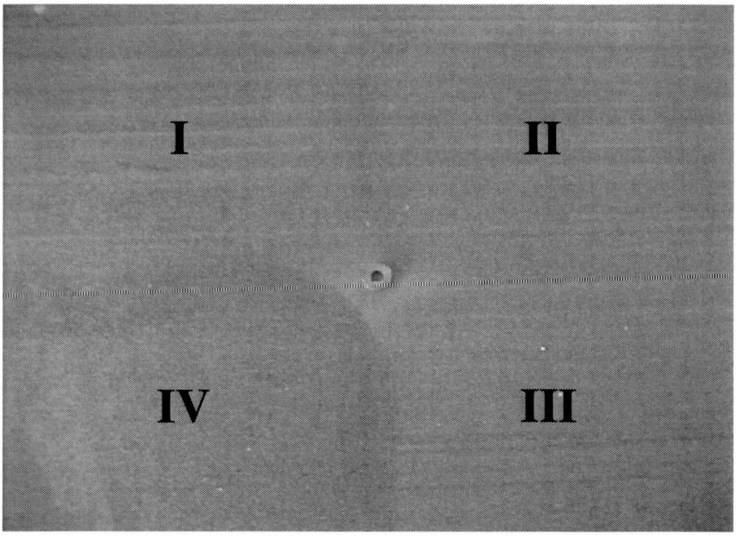

Figure 4. Granite plate after the grinding step performed by the first 4 of crowns: I – group 1; II – group 2; III – group 3 and IV – group 4).

Figure 5. (a) and (b) Details of an array of 6 grinding crowns mounted in the industrial grinding machine satellite. Each crown rotates at 600 RPM under a load of 500 N.

3. Results and Discussion

3.1. Small Samples

3.1.1. Surface Finishing (mean Ra) and Abrasion Resistance

Table 1 shows the results of surface finishing (SF) and weight loss (WL) for the epoxy - quartz and the SOREL–20 % weight SiC composites, calculated according to Equations 1 and 2. The presented data are quite reproducible and repeatable, once the mean values correspond to seven specimens, and the standard deviations were too small.

Surface finishing values of most epoxy-quartz composites (except epoxy-70 % quartz) were superior to the commercial crown SOREL- 20% SiC. Epoxy–quartz composites with 80 and 85 % quartz showed the better surface finishing results. This was due to the higher volume of abrasive particles of these samples when compared to the composites containing 70 % and 75 % quartz and the SOREL- 20% SiC. Specimen with 90 % silica showed a considerable drop in surface finishing. This was attributed to the fact that a small contact area enables to bond the abrasive particles during the abrasion process. According to Przyklenk [3], an increase in the content of abrasive particles into the composite matrix (in this case, quartz) leads to an increase in the surface finishing up to a limit, after which it begins to fall, just because the matrix adhesive area is diminished, and abrasive particles pull-out easily.

Weight loss values of all epoxy-quartz composites were at least slightly superior (better abrasion resistance) to the commercial crown SOREL- 20% SiC. Despite of the higher hardness of SiC, when compared with SiO_2, the epoxy matrix promotes a higher adhesion to ceramic particles than the magnesian cement (SOREL) [2, 3].

Relative weight loss was calculated dividing weight loss of the SOREL crown by the weight loss of each composite epoxy-quartz, and it was shown at the right column of Table 1. When compared to the SOREL- 20% SiC, composites containing 80 and 85 % quartz showed a relative weight loss of 8.6 and 34.6 times inferior, respectively. Results in Table 1 also show that the 80 and 85 % epoxy – quartz composite offers better surface finishing (SF) to the ornamental rock and a reduced tool wear, which means a higher tool useful life. The amount of quartz of 70 % is not enough to ensure a good performance in a tool – it does not protect the matrix by the abrasion wear during operation, so that just a few contact points between the stone and the abrasive particles occur, consequently the stone surface finishing is inferior and the wear (weight loss) is high. On the other hand, 90 % quartz reduces dramatically the adhesion area between the matrix and abrasive particles, so that pull-out takes place prematurely. These results are in agreement with [3], and show that the 85 % quartz – epoxy is the best processed composite in this study.

Table 1. Surface finishing (SF), weight loss (WL) and relative weight loss (Relative WL) for the composites tested. Results showed refer to the mean values obtained

Composite	SF (mm)	WL (g)	Relative WL
Epoxy- 70 % quartz	0.018 ± 0.005	0.785 ± 0.008	1.103
Epoxy- 75 % quartz	0.055 ± 0.008	0.105 ± 0.011	8.281
Epoxy- 80 % quartz	0.149 ± 0.007	0.100 ± 0.006	8.663
Epoxy- 85 % quartz	0.126 ± 0.004	0.025 ± 0.006	34.642
Epoxy- 90 % quartz	0.080 ± 0.006	0.129 ± 0.008	6.713
SOREL- 20 % SiC	0.032 ± 0.007	0.866 ± 0.010	1

3.1.2. Wear Surface Microstructures

The surface microstructure of the specimen epoxy-70 % quartz is shown in Figure 6a, where the significantly worn epoxy resin matrix can be observed. This was a consequence of its friction with the irregular surface of the granite disc. Just a few points of cut of the SiO_2 were able to cut the stone (smaller amount of abrasive particles), so the matrix suffered the observed wear, and the "holes" (center, top center, and left) in this figure are regarded to the pull-out of SiO_2 particles. Figure 6b shows that the worn surface microstructure of the specimen epoxy-75 % quartz presents a reduced wear compared to the specimen epoxy-70 % quartz, showing an increased number of SiO_2 particles still bonded in the matrix, although a hole is observed at the left bottom and right top of the figure. Figure 6c shows that the surface microstructure of the specimen epoxy-80 % quartz also presented a reduced wear compared with the specimen epoxy-70 % quartz and 75 % quartz, also where the epoxy matrix promoted a higher adhesion of the abrasive particles. It also shows the detachment of particles (pull-out in the center) due to the abrasive process and new quartz particles arising in the surface (left side) which promoted a continued grinding action. Figure 6d (epoxy-85 % quartz) shows a homogenous distribution of silica particles in the epoxy matrix, with a guaranteed retention of abrasive particles during the test. This enabled the epoxy – 85 % quartz the lower weight loss value, when compared to the all other small samples, as observed in Table 1. Figures 6a-d show rough surfaces which indicate the occurrence of ductile wear, as stated in [6-9].

The surface microstructure of the sample epoxy-90 % quartz is shown in Figure 6e, where flat regions are seen, in contrast with all other SiO_2 samples. It takes place because the large amount of SiO_2 particles which pull-out remains in the contact zone between the worn matrix and the stone, thus abrading the matrix in such a way that this latter presents a flat surface, once the loose SiO_2 particles cause a polishing on it. Facing such situation, the matrix wear process and SiO_2 pull-out become competitive, and matrix flattening along with slipping between the surface composite and the granite takes place. As consequence, this sample presented the lower overall performance of all specimens of the system epoxy-quartz.

Sample SOREL-SiC also showed a degraded surface, as can be seen in Figure 6f. Adhesion of the matrix was not adequate because just a few abrasive particles can be seen present in the matrix. This is in agreement with the higher values of weight loss (lower abrasion resistance) observed for this specimen when compared to the epoxy-quartz samples. In addition, the wear surface shows that its degradation happens through the releasing of fine grits from the matrix, which characterize a condition of brittle wear (according with [6-9]), and consequently the surface becomes flat. This was expected due to the ceramic nature of the matrix (brittle), with different behavior than the polymeric matrix (ductile).

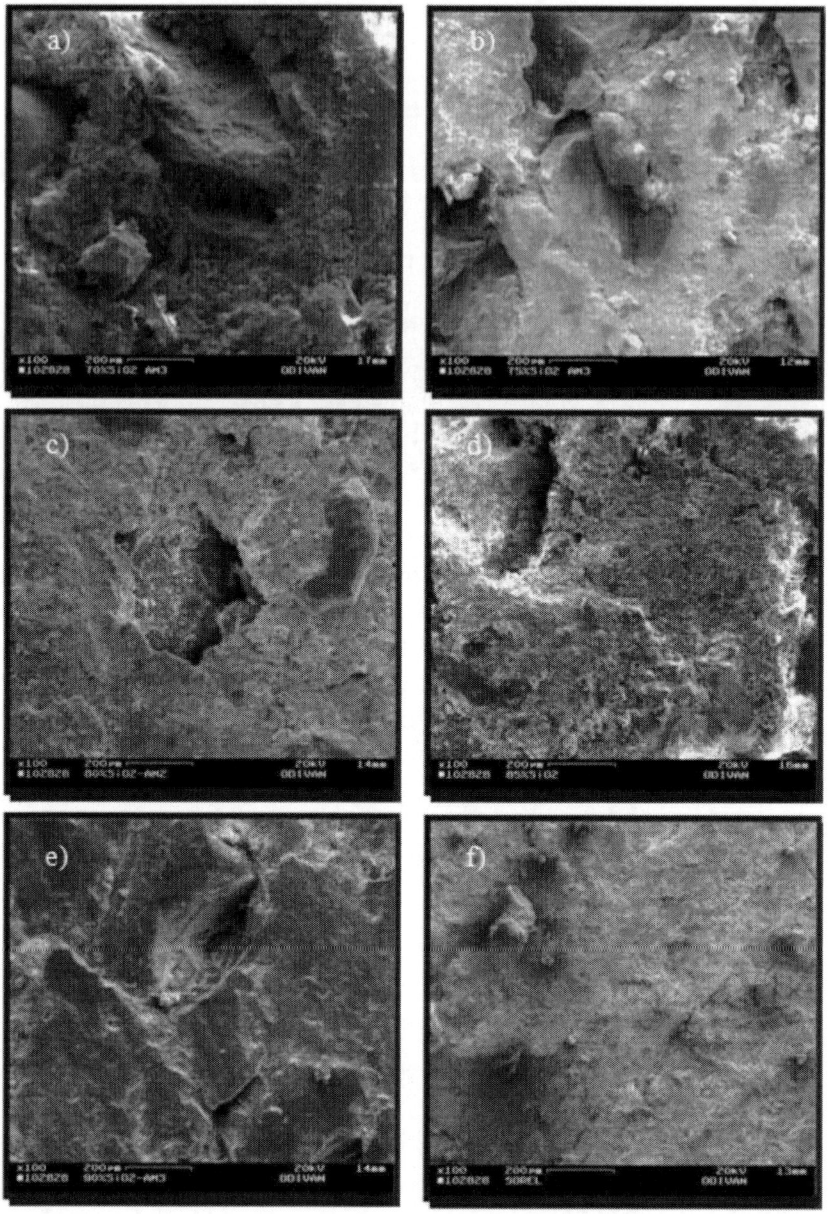

Figure 6. Surface of the small samples: (a) epoxy-70 % quartz; (b) epoxy–75 % quartz; (c) epoxy-80 % quartz; (d) epoxy-85 % quartz; (e) epoxy-90 % quartz; (f) SOREL-20 % SiC.

3.1.3. Comments about the Abrasive Composites Wear Mechanism

For the two classes of composites studied, epoxy-quartz (ductile matrix and high hardness abrasive) and SOREL-SiC (brittle matrix and high hardness abrasive) to be used for the grinding of ornamental rocks, the following wear mechanism was observed:

1) initial abrasion of the agglomerated matrix;
2) beginning of the stone cut by the abrasive particles;
3) dulling of the abrasive particles up to their flattening and ultimate detachment from the matrix – when a satisfactory adherence matrix-abrasive particles is observed (ductile matrix – high hardness particles), see Figure 6d. For the case of a bad adherence, particles at the beginning of the grinding suffer detachment causing a premature and excessive wear of the matrix (brittle matrix - high hardness particles), as seen in Figure 6f;
4) after detachment, further wear of the polymeric matrix allow new cutting faces of other particles that arise at the surface, restarting the cycle.

For the above described wear mechanism, the abrasive particles content in the composite plays an important role. When this content is increased, the tool useful life is increased up to a determined composites' stoichiometric value. Beyond this limit, a reduction in the tool properties is observed, due to the reduction of matrix useful area for the anchorage of the abrasive particles. The adherence matrix-particle is considerably reduced because the adhesion strength is directly proportional to the anchorage area, leaving particles to be prematurely detached.

From the point of view of the composites‘ wear mechanism, and looking at Figures 6a-f, there are no indications of the self-sharpening process of the abrasive particles mentioned by Wright and Wapler [6] and Davis [7], but it is in fair agreement with Warnecke and Wimmler [10] when studying the polishing of advanced ceramics. The grinding of ornamental rocks involves the combined action of vertical load and rotation movement imposed to the tool, what does not happens for the cut operation, becoming the wearing process for this case more dramatic than the situation described by Wright and Wapler [6] and Davis [7].

3.1.4. Compression Strength

Figure 7 shows the graph load (N) versus time (s) for specimens containing 70, 75, 80, 85 and 90 % weight of quartz and SOREL-SiC. The specimen with 70 % quartz presented the highest compressive strength among all quartz composites. This is due to the higher amount of epoxy resin of this sample, offering higher resistance to the applied load. The polymer used is considered to show considerable high compression strength, and during its cure, porosity in the matrix was not observed.

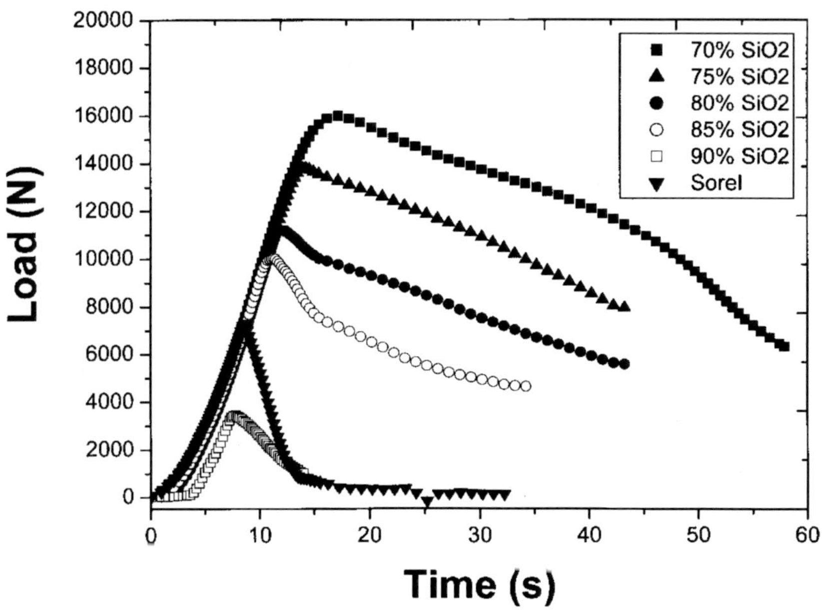

Figure 7. Compressive load versus time for all tested composites.

Although SOREL-SiC is a ceramic material, its compression strength was twice as high as the composite with 90% quartz, due to the increased content of quartz particles present in this specimen. Large amounts of abrasive particles embedded into a polymeric matrix diminish the compressive resistant area, as predicted in [2, 6]. Despite this, it was stated earlier that the industrial load applied to the commercial abrasive crowns is 500 N, which indicates that all processed composites can withstand the industrial compressive conditions once the worst result is 5 times superior to this industrial condition.

3.2. Grinding Crowns

The composites epoxy-(75-80-85) % quartz presented the best global results of WL, SF and compressive strength, as seen previously for the small samples, so they were used to process the grinding crowns.

3.2.1. Height, Volume and Mass Loss

Table 2 evidences the mean height (initial; intermediary – 90 seconds test; and final – 180 seconds test) for each group of grinding crown (1 to 4) used in the grinding granite in the stone industry. It is seen that all groups experience reduction in height, as expected.

Table 3 shows the results of height reduction for all groups of crowns, for 90 and 180 seconds of grinding tests. It is evident that the system SOREL-SiC presented the highest height loss among all crowns, *i.e.*, 6.5 times higher than the epoxy-quartz crown. It means that the later is more wear resistant.

As said previously, according to Table 3, it is possible to observe that the height reduction caused by the crowns' wear is much higher for group 4, which corresponds to the SOREL-SiC crowns – commercial ones. The wear for the epoxy-75 % quartz crowns experiences a wear until 28 times inferior than the SOREL-SiC crowns.

Table 4 shows the results of mass loss for each crown group after a 3 minute grinding test. It can be noted that group 1 crowns had no significant wear, and it can negatively affect the grinding surface quality of the granite plate, as will be seen in the following.

Epoxy-quartz based crowns of groups 2 and 3 presented wear of 8.9 and 3.9 times inferior than the SOREL-SiC crown, respectively. These results are of great relevance, once the processed crowns present a longer lifetime. Despite of these values, the final roughness and microstructure generated in the granite plate must be considered, as predicted by Aigueira et al [11].

3.2.2. Microstructures of the Granite Plate

The analysis of optical microscopy reveals the quality of the granite plate surface before and after the grinding process with the crowns groups 1-4 – see Figures 8-12.

Table 2. Mean heights of the crowns for the initial stage – prior to the grinding test, at the intermediary stage – 90 seconds test, and final stage – 180 seconds test

Group	Mean initial height (cm)	Mean Intermediary height (cm)	Mean final height (cm)
1	39.46	39.29	39.14
2	39.81	39.40	39.17
3	31.70	30.94	30.31
4	51.61	47.04	42.61

Table 3. Height reduction for the 4 groups of crowns

Group	Intermediary height reduction (cm)	Final height reduction (cm)	Total height reduction (cm)
1	0.17	0.15	0.32
2	0.41	0.23	0.64
3	0.76	0.63	1.39
4	4.57	4.43	9.00

Table 4. Initial and final mass, along with the mass loss ($\Delta m = m_i - m_f$), and mass loss percentage ($\Delta m/m_i$) for each crown group. m_i and m_f are, respectively, the initial and final mass of each crown

Group	Initial mass (g)	Final mass (g)	Mass loss (g)	% mass loss
1	900	896	4	0.4
2	950	940	10	1.1
3	1000	976	24	2.5
4	1820	1658	162	9.8

For the granite plate surface prior to the grinding step – Figure 8, one can observe a uniform microstructure with some pores and too rough. This is a common microstructure of a stone plate that was submited to the cutting (slicing) process by linear saws or diamond wires [12].

The granite surfaces grinded by crown groups 1-3 are depicted by Figures 9-11. It is very clear that an increase in the quartz content improves the grinding operation, revealing, progressively, the grains and its boundaries and microconstituents of the stone. The surface is also progressively improved in the point of view of roughness, as seen by the absence of ondulations in Figure 11.

Figure 12 is regarded to the granite surface grinded by the SOREL-SiC crown. The microstruture is similar, but slightly better than that from crown 3 (epoxy-85 %wt quartz). One can observe a better appearance of the grains of the granite plate. Nevertheless, it leaves on the plate small pieces of oxidized iron (tough reinforcement) – as seen by a shadow at the top center of Figure 12. The iron is a constituent of the SOREL-SiC based crown, it is put in a fraction of about 5%wt to ease the lubrication at the contact zone during the grinding process, but it is also deleterious because it reduces the microstructure quality of the grinded stone, as can be observed.

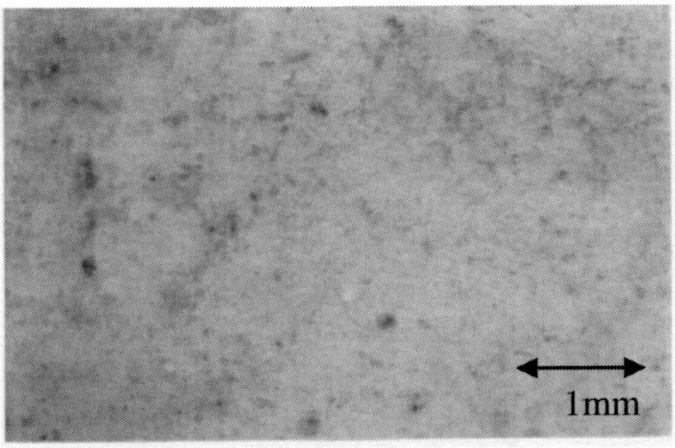

Figure 8. Microstructure of the granite plate before the grinding process. 50X magnification.

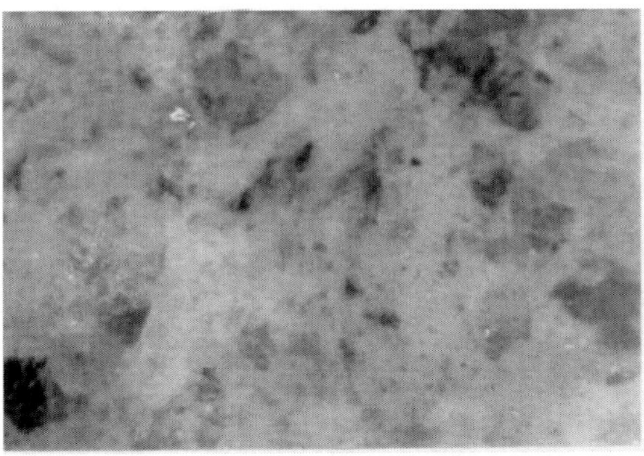

Figure 9. Microstructure of the granite plate after grinding by the group 1 crown. 50X magnification.

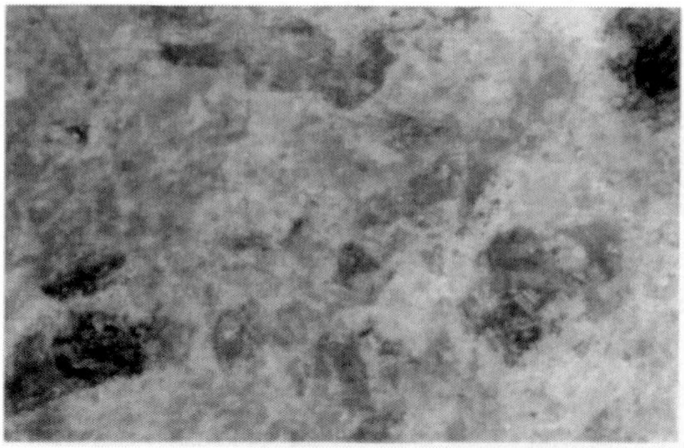

Figure 10. Microstructure of the granite plate after grinding by the group 2 crown. 50X magnification.

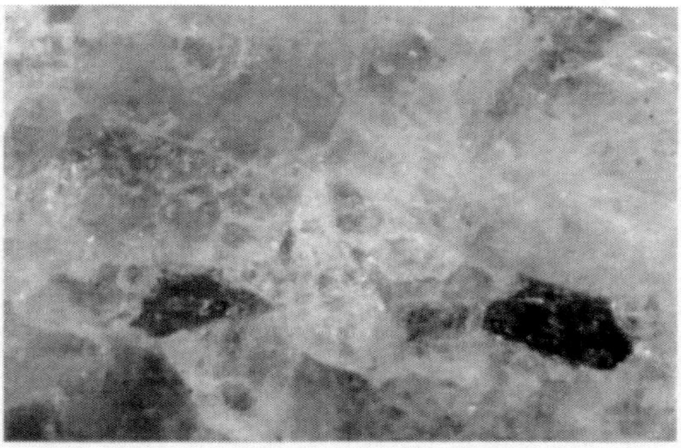

Figure 11. Microstructure of the granite plate after grinding by the group 3 crown. 50X magnification

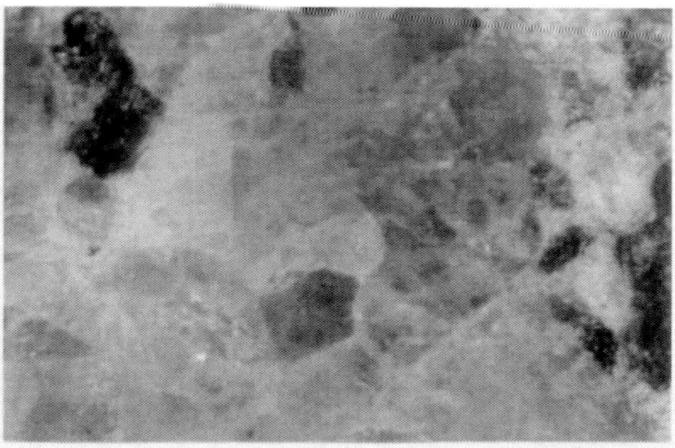

Figure 12. Microstructure of the granite plate after grinding by the group 4 crown. 50X magnification

3.3.3. Surface Finishing (mean Ra) of the Granite Plate

Table 5 shows the mean Ra results for the granite surface grinded by the 4 groups of crowns. It is observed that the crowns epoxy-75 %wt quartz and epoxy-80 %wt quartz do not improve significantly the surface quality of the granite plate, considering as 0.022 mm the roughness of the granite plate prior to the grinding process. It corroborates with the observed microstructures presented in Figures 4-6.

Table 5. Surface finishing results for the granite plate prior to the grinding step (0), grinded by the epoxy-75 %wt quartz crown (1), grinded by the epoxy-80 %wt quartz crown (2), grinded by the epoxy-85 %wt quartz crown (3), grinded by the SOREL-SiC crown (4)

Granite plate surface	0	1	2	3	4
mean Ra, mm	0.022 ± 0.006	0.018 ± 0.007	0.012 ± 0.006	0.005 ± 0.003	0.004 ± 0.002

On the other hand, the surface quality obtained when using an epoxy-85 %wt quartz crown is equal to that from the SOREL-quartz crown, as expected, due to the increased quartz amount, which improves the granite surface conditions, and it is seen in Figure 7 and 8. These results are in agreement with [14, 15].

CONCLUSION

The following main conclusions can be drawn from this exploiting work. It was divided in two groups.

A) Regarding the small samples:

1) The developed route for the processing of composites based on the system epoxy-quartz to be applied as abrasive crowns for grinding of ornamental rocks was attained with success, being made fast and with simplicity;
2) Among the composites of the epoxy-quartz system, the composition epoxy-85 % quartz was considered to present the best results. Surface finishing, compression strength and wear

resistance were considerably above to the commercial SOREL-20 % SiC composite;

3) The self-sharpening of the abrasive particles was not observed during the wear mechanism of the composites studied. The ductile behavior of the epoxy matrix dominated the wear mechanism. A satisfactory adherence of the quartz particles in the epoxy matrix was observed.

B) Regarding the grinding crowns:

1) The obtained results during the dimensional and mass loss tests suggest that the epoxy-85 %wt quartz crowns showed the best results when compared to the commercial SOREL-SiC crowns;
2) The microstructure of the granite plate grinded with the epoxy-85 %wt quartz is close to that of the SOREL-SiC crown;
3) Through the roughness tests it was possible to verify that the surface quality produced by the epoxy-85 %wt quartz is as good as that from the SOREL-SiC crown;
4) As an additional result, it is worth to emphasize the processing time economy to manufacture the crowns because while the SOREL-SiC crowns take about 20 days curing, the epoxy-quartz demands just 5 hours, resulting in an improved productivity allied to a competitive quality.

ACKNOWLEDGMENTS

The author expresses his gratitude to the São Salvador stone processing company for the grinding tests facilities. He also thanks to the Brazilian Agency CNPq, and to the State University of North Fluminense – UENF.

REFERENCES

[1] Xu, X. (2001). Experimental study on temperatures and energy partition at the diamond-granite interface in grinding. *Tribology International*, 34, 419-426.

[2] Aigueira, R.B., Bobrovnitchii, G.S., Filgueira, M. (2004). Processamento de um compósito abrasivo à base do sistema SiC-β –

poliéster para polimento em rochas ornamentais. *Anais do* 59º Congresso Anual da ABM, 1527-1535.

[3] Przyklenk, K. (1993). Diamond impregnated tools – uses and production. IDR – *Toolmaking*, 4, 192-195.

[4] Tolkowsky, M. (1920). Research on the Abrading, Grinding or Polishing of Diamond. D. Sc. Thesis, University of London.

[5] Wilks, J., Wilks, E. M. (1979). Abrasion and wear of diamond. In: J.E. Field (ed.), Properties of Diamond, Academic Press, London, 351 – 381.

[6] Wright, D. N., Wapler, H. (1986). Investigations and prediction of diamond wear when sawing. *Annals of the CIRP*, 35 1, 239-244.

[7] Davis, P. R. (1996). An indicator system for saw grit. IDR, 3, 78-87.

[8] Lee, G. Y. (2002). A physically-based abrasive wear model for composite materials. *Wear*, 252, 322-331.

[9] Luo, S. Y. (1999). Effect of fillers of resin-bonded composites on diamond retention and wear behavior. Wear, 236, 339-349.

[10] Warnecke, G., Wimmer, J. (1995). Stock removal and wear in deep grinding high-performance ceramics. *IDR*, 3, 126-132.

[11] Aigueira, R. B. et al. (2005). Processamento e avaliação de compósito diamantado para o polimento de rochas abrasivas. *Matéria*, 10 3, 419-428.

[12] de Oliveira, O.C. et al. (2008). Processamento e desgaste abrasivo de compósitos à base de epóxi-sílica para uso em coroas de polimento de rochas ornamentais. *Matéria*, 13 1, 88-95.

[13] Aigueira, R. B., Filgueira, M. (2006). Mecanismo e resistência à abrasão de compósitos à base de poliéster-SiC para uso em coroas de polimento de rochas ornamentais. Polímeros: *Ciência e Tecnologia*, 16 3, 187-192.

[14] Aigueira, R. B. et al. (2006). Characterisation of composites based on the system polyester-SiC for use in the polishing of ornamental rocks. *Int.J. Machining and Machinability of Materials*, 1 1, 62-75

[15] Azeredo, S. R. et al. (2008). Desenvolvimento de um novo compósito abrasivo de desbaste de rochas ornamentais. *Matéria*, 13 1, 203-208.

In: Recent Advances in Abrasives Research
Editor: Dirk Bähre
ISBN: 978-1-62808-566-2

Chapter 2

ABRASIVES UTILIZED IN FRICTION COMPOSITES FOR TAILORED BRAKE LINING APPLICATIONS

Shicheng Qi[1], Rongping Yun[2], Vlastimil Matejka[3], Shengling Jiang[1], Xiaoa Zhang[1] and Yafei Lu[1,1]

[1]Key Laboratory of Carbon Fiber and Functional Polymers, Ministry of Education, Beijing University of Chemical Technology, Beijing, China

[2]High-Tech Research Institute, Beijing University of Chemical Technology, Beijing, China

[3]Nanotechnology Centre, VSB-Technical University of Ostrava, Ostrava-Poruba, Czech Republic

ABSTRACT

Abrasives in brake friction materials play a critical role in friction and wear performance. In this article, factors (such as Mohs hardness, volume fraction, particle sizes of abrasives and tribochemistry) affecting friction coefficient, wear rate, formation of friction layers and disc life were discussed. The critical volume fraction of abrasives was experimentally determined and explained by a competitive relation between adhesion and abrasion friction mechanisms. The effect of the

---

[1] Corresponding author. Email: ylu623@gmail.com

particle size of abrasives on friction and wear performance can be explained by abrasion friction and wear model where two parameters, *r* (radius of abrasives) and *d* (depth of abrasives embedded into disc), are involved. The area of friction layers caused by abrasives including primary and secondary contact plateaus can be calculated by using an elliptical model. Formation of friction layers was also simulated by a two/three-dimensional cellular automata model.

Keywords: Friction, wear mechanisms, composites, brakes

INTRODUCTION

Phenolic matrix brake friction composites called as organic friction materials are multiple component composites, which usually contain four classes of ingredients: abrasives, solid lubricants, fiber reinforcements and space fillers [1, 2]. Modern friction composites can be roughly divided into two groups, semi-/low metallic and Non-asbestos organic (NAO), as shown in Figure 1. Semi- and low metallic friction composites are composed of steel wool or copper and brass chips as reinforcements and contain iron powders, graphite, metal sulfides and other fillers [3-5]. Non-asbestos organic (NAO) friction composites are composed of aramid fibers or natural fibers as reinforcements and contain potassium titanate, wollastonite, vermiculite, barite, graphite and other fillers [5, 6].

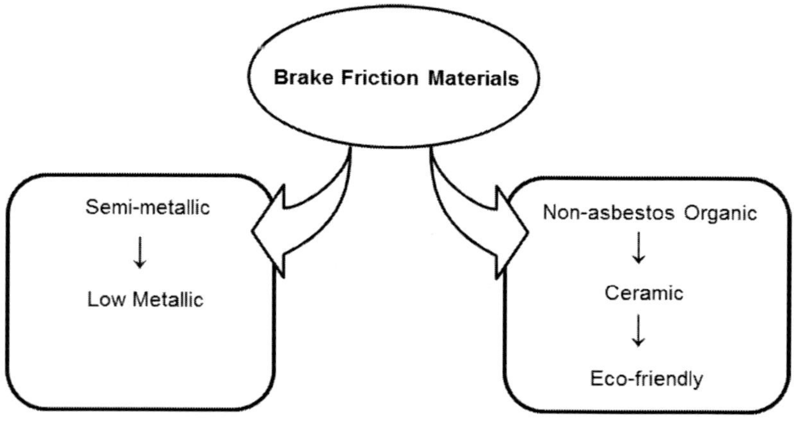

Figure 1. Modern Brake Friction Composites.

The friction coefficient (μ) is often used as measure of resistance against the relative motion of objects due to the existence of adhesion and abrasive phenomenon, generally defined as:

$$\mu = f/N \quad (1)$$

where f is the friction force, while N is the normal force.

Brake friction composites require a higher friction coefficient at elevated temperature, as well as a relatively stable μ under different operation conditions during braking processes. Therefore, not only a high value of μ is important, but friction stability is considered even as a more crucial parameter affecting the final friction performance. Wear is related to the interaction between surfaces of friction couples, which causes mass or volume losses. A commercial brake friction composite requires a properly low wear rate for a long lifetime.

Due to the non-linear and complex relationship between composition and friction performance (such as friction coefficient (μ) and wear (ω)), we have developed a combinatorial friction materials research (CFMR) approach to screen raw materials for the development of friction composites, as well as to optimize the related friction and wear performance [7-11]. Five critical ingredients were selected to formulate and prepare a series of model friction composites by applying the Latin square design method. In the model samples, graphite served as lubricant, aramid pulp (Twaron) as organic reinforcement, steel wool as metal reinforcement, alumina Al_2O_3 as abrasive, and phenolic resin as binder. The friction composite samples were prepared in several steps: blending, hot-press molding, thermal treating, and post-processing. Then, the proper friction test rig was chosen to assess the friction performance of the prepared composites. Based on the evaluated effect of each ingredient on friction and wear performance, all five ingredients were ranked as follows [12]:

$$Al_2O_3 > \text{Twaron} > \text{Graphite} > \text{Steel Wool} > \text{Phenolic Resin} \quad (2)$$

This result indicates that the abrasive is the most important ingredient among all five ingredients, mainly due to its ability for enhancing and stabilizing the friction coefficient of friction composites. The abrasive used in a brake friction material requires a higher Mohs hardness (≥ 7) than that of its counterpart disc made of cast iron (= 5). The candidate abrasives with relatively high Mohs hardness are listed in Table 1, while ceramic fibers, metals and their oxides usually used in friction composites are also mentioned in this chapter.

Abrasion and adhesion are two main mechanisms occurring in a braking process [13, 14]. In general, without abrasives, the friction interface shows the adhesion friction mechanism. It results in a relatively low friction coefficient, but a long life of friction composite. Addition of abrasives, the friction interface shows the abrasion friction mechanism. It results in a relatively higher μ, but also a higher specific wear rate. Abrasives can also damage the disc made of cast iron during braking processes. The factors affecting the friction performance mainly include Mohs hardness, the size of particles and the content of abrasives. The general advantage of abrasives is on the one hand to enhance μ even at an elevated temperature, on the other hand they increase the wear of the friction composite as well as the disc. With respect to service requirements given on the brake linings, the abrasives have to be used in brake friction composites, but their volume fraction has to be limited to usually less than 0.10 [15, 16].

Table 1. Mohs Hardness List of Abrasives, Ceramic Fibers and Metals Used in Friction Composites

Abrasives, Ceramic Fibers and Metals	Mohs Hardness
Silicon Carbide (SiC)	9
Alumina (Al_2O_3)	7-8
Zircon ($ZrSiO_4$)	7-8
Silica (SiO_2)	7
Zirconia (ZrO_2)	7
Ferroferric Oxide (Fe_3O_4)	6-7
Ferric Oxide (Fe_2O_3)	5
Steel Wool	5
Iron Powder	4-5
Wollastonite ($CaSiO_3$)	4-5
Basalt Fiber	5-9
Glass Fiber	5.5-6.5
Cast Iron (Disc)	5

Abrasives for brake friction lining application usually are mainly inorganic materials in particulate or fiber states, such as $ZrSiO_4$, ZrO_2, Al_2O_3, SiO_2, and SiC. The effects of ceramic particulates and fibers on the friction performance of brake friction composites were studied by some researchers in the past few years [17-20]. Our research group has also finished some investigation about the effects of three abrasives (SiC, Al_2O_3 and $ZrSiO_4$) on friction and wear performance of friction composites in both semi-metallic and NAO (ceramic and eco-friendly) types [21-27]. The factors related to the abrasives affecting friction performance are multiple. In this chapter, we will mainly focus on several important aspects about the role of the abrasives in brake friction composites.

MEASUREMENTS AND EVALUATION OF FRICTION PERFORMANCE

Friction coefficients and specific wear rates of brake friction composites were obtained during and after measuring by friction tests performed under constant load and constant speed of the cast iron disc. The friction tester used in these experiments was also equipped with a temperature control system through heating and cooling devices, in order to fulfill test procedures according to China National Standard GB 5763-2008 (Table 2). The μ values at different temperatures are the main part of outputs from a whole friction test. The friction coefficient during the heating process, assigned as μ_f, can be measured from 100°C to 350°C at an interval of each 50°C, while the friction coefficient during the cooling process, assigned as μ_r, can be obtained in a similar way from 300°C to 100°C.

Table 2. Allowed Values of Friction Performance of Brake Friction Materials (GB 5763-2008)

	100°C	150°C	200°C	250°C	300°C	350°C
μ	0.25-0.65	0.25-0.70	0.25-0.70	0.25-0.70	0.25-0.70	0.25-0.70
ω	0-0.50	0-0.70	0-1.00	0-1.50	0-2.00	0-2.50

The average friction coefficient (μ_a) is defined by:

$$\mu_\alpha = \frac{\Sigma_{i=1}^{n} \mu_{fi}}{n} \quad (3)$$

where $n = 6$.

The friction stability ($\Delta\mu^*$) is defined as follows:

$$\Delta\mu^* = \frac{\sum_{i=1}^{n-1}\left(N_i x \frac{|\mu_{fi}-\mu_{ri}|max - |\mu_{fi}-\mu_{ri}|}{|\mu_{fi}-\mu_{ri}|max - |\mu_{fi}-\mu_{ri}|min}\right) + N_n}{n} \quad (4)$$

where $i = 1$ (100°C), 2 (150°C), 3 (200°C), 4 (250°C) and 5 (300°C), respectively. N_i is the weight at i temperature and selected from equation (4), while N_6 is a weight at 350°C that can be directly obtained from experimental results (the value of N_6 is numerically equal to μ_f measured at 350°C). According to the Chinese standard for automotive friction materials (GB5763-2008), both μ_f and μ_r values at a series of different certain temperatures should be in a limited interval, 0.25-0.70 for 150, 200, 250, 300 and 350°C, except 0.25-0.65 for 100°C.

$$N = \begin{cases} 1.00, & \mu \in [0.40,0.50] \\ 0.75, & \mu \in [(0.35,0.40) \cup (0.50,0.55)] \\ 0.50, & \mu \in [(0.30,0.35) \cup (0.55,0.60)] \\ 0.25, & \mu \in [(0.25,0.30) \cup (0.60,0.70)] \\ 0.00, & \mu \in [(0.00,0.25) \cup (0.7, X\ )] \end{cases} \quad (5)$$

The specific wear rate ($\omega(T_j)$) can be calculated with the following formula:

$$\omega(T_j) = \frac{1}{2\pi R} \times \frac{A}{n'} \times \frac{d_1 - d_2}{f_m} \quad (6)$$

where j = 1 (100°C), 2 (150°C), 3 (200°C), 4 (250°C), 5 (300°C) and 6 (350°C), respectively, d_1 and d_2 are the average thickness (cm) before and after tests, respectively, R is the distance between the center of the sample to the center of the disc (= 0.15 m), n' is the rotating number of the disc (= 5000), A is the surface area of the tested sample (= 2.5×2.5 cm^2) and f_m is the average friction force (N).

The average specific wear rate (ω_a) is defined by:

$$\omega_a = \frac{\sum_{j=1}^{n} \omega_j}{n} \quad (7)$$

Characterization Methods

The morphology of the abrasives and friction surface of the tested samples was observed by scanning electron microscope (SEM) Philips XL30 with energy dispersive X-ray microanalysis (EDX). The SEM images were obtained using back scattered electrons at operating voltage of 25 kV.

Effects of Abrasives on the Friction Coefficient

Effects of Hardness on the Friction Coefficient

The hardness of an abrasive plays a key role in determining the friction coefficient of a friction composite. Table 3 shows the effects of the three types of abrasives (SiC, Al_2O_3 and $ZrSiO_4$) with different hardness on μ of ceramic friction composites [20]. Ceramic friction composites were composed of potassium titanate (K_2TiO_6), wollastonite ($CaSiO_3$), aramid pulp (Twaron), brass chips, baryte ($BaSO_4$), nitrile rubber (NBR), phenolic resin and 5.6 vol. % of the given abrasive. The average particle sizes in diameter of these three abrasives are in the range 1-10 μm. The μ value of the sample without any abrasive added is significantly lower, when compared with the other three samples. Upon adding abrasives into friction composites, μ is enhanced with the increasing Mohs hardness, which has been verified in practice. Therefore, hardness is an important factor on affecting μ values, which is mainly manifested as a positive contribution to the enhancement of μ values. According to the requirements of μ values, the abrasive with a suitable hardness can be selected for designing brake friction formulations.

Table 3. Effects of Hardness of Abrasives on μ_a

	SiC	Al_2O_3	$ZrSiO_4$	None
μ_a	0.600	0.536	0.465	0.322

What deserves to be mentioned is, during braking processes, tribochemical reactions will lead to generating some new kinds of abrasives in situ. Therefore, the content and Mohs hardness of abrasives in friction composites will be changed during the friction process. For example, iron in the surface of friction composites, existent in steel wool, iron powder, or transferred matter from the counterpart, can be oxidized into Fe_3O_4 and Fe_2O_3. Besides, metal sulfide such as Sb_2S_3 with a Mohs hardness of 2.5 can be oxidized into a harder metal oxide, Sb_2O_3 (Mohs hardness = 5.2).

Effects of Particle Sizes on the Friction Coefficient

The particle size of an abrasive is another important factor affecting the μ values of friction composites. However, little has been reported about the effects of the particle size and its distribution of abrasives on μ and ω in the past few years [28-30]. We have finished some studies on the effects of the morphology and particle sizes on friction performance, using three different SiC abrasive particles (F360-40 μm, F600-10 μm and F1200-3 μm) as shown in Figure 2. For ceramic friction composites, the effect of the SiC particle size on μ_a is shown in Table 4. The results indicate that the composites with addition of the finest SiC particles show higher μ_a values, almost regardless of SiC contents. Furthermore, friction composites with the finest SiC particles added show the best friction stability [30].

Figure 2. Morphology and Average Particle Sizes of SiC: (a) F360 (40 μm), (b) F600 (10 μm) and (c) F1200 (3 μm).

Table 4. Effect of Size of SiC Particles and Their Content (0, 3.4, 5.6, 9.0 and 14.6 vol. %) on μ_a

	SiC-0	SiC-3.4	SiC-5.6	SiC-9.0	SiC-14.6
3 μm	0.281	0.432	0.447	0.449	0.460
10 μm	0.281	0.411	0.392	0.426	0.438
40 μm	0.281	0.394	0.402	0.408	0.472

In order to understand the effects of the particle size of abrasives on μ of the friction composites, an abrasion friction model was proposed as shown in Figure 3, in which two parameters, *r* (radius of abrasives) and *d* (depth of abrasives embedded into the cast iron disc), are involved. The *d* value increased with the *r* value increasing under the same normal force loaded. In addition, the abrasive with a larger particle size can result in a higher level of abrasion.

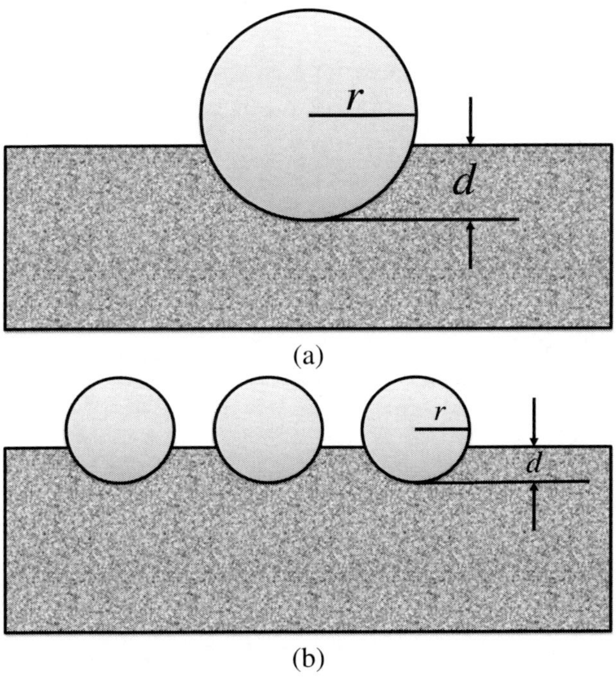

Figure 3. Abrasion Friction and Wear Model: (a) Large Particle Size and (b) Small Particle Size.

Effects of the Content of Abrasives on Friction Coefficient and Critical Volume Fraction of Abrasives

The effect of the abrasive content in organic friction composites on μ_a is clearly observed in Table 4 [30]. The results indicate that μ_a values enlarge with the SiC-F360 (40μm) content increasing.

As described above, μ of the friction composites without abrasives seems lower when compared with μ of samples with abrasives. It is also found that the critical volume fraction (V_c) of abrasives properly exists and falls in the interval 0.034-0.056 (Table 5 [31]). The values of μ rise rapidly along with the abrasive content until V_c is reached, and then just increase slightly, or even decrease with the abrasive content increasing further. To explain this phenomenon, we presume that two friction mechanisms, adhesion and abrasion, compete during the friction process. The adhesive friction effect is dominant in friction composites without abrasives, while the abrasive friction mechanism mainly occurs during the friction process of the samples with abrasives. Both friction mechanisms occur simultaneously during a friction process, but their relative proportion changes with the abrasive content increasing. Consequently, the abrasive mechanism maybe plays a dominant role for the composite with a high abrasive content.

Table 5. Friction Stability of Semi-metallic Composites Containing $ZrSiO_4$

	Zr-0	Zr-3.4	Zr-5.6	Zr-9.0
Δμ*	0.32	0.37	0.60	0.45

Effects of Abrasives on Wear

The specific wear rate is one of the most important parameters related to the lifetime of a friction composite. As for wear of friction materials, there are mainly two types: the abrasive wear and the adhesive wear [32]. Abrasives generally play a negative role in depressing of the specific wear rate. Usually, the adhesion friction mechanism is responsible for a lower wear, contrary to the abrasion friction mechanism. Table 6 shows a comparative result of specific wear rates of friction composites without and with 5.6 vol. % abrasive. In addition to the given abrasive, the composites also contain several other ingredients, including K_2TiO_6, $CaSiO_3$, aramid pulp (Twaron), brass chips, $BaSO_4$, nitrile-butadiene rubber (NBR) powder, and phenolic resin [20]. The effects of the SiC particle size and its content on ω_a are summarized in Table 7, which indicates that the ω_a values of friction composites rise along with the SiC amount. Moreover, the addition of SiC particles in a larger size could improve the wear resistance of the friction composites, which shows lower values of ω_a during the friction tests.

Table 6. Effects of Abrasives with Different Hardness on ω_a

	SiC-5.6	Al_2O_3-5.6	$ZrSiO_4$-5.6	Abr-0
ω_a, $\times10^{-7}cm^3/Nm$	0.646	0.570	0.361	0.381

Table 7. Effects of Particle Sizes and Content of SiC on ω_a

ω_a, $\times10^{-7}cm^3/Nm$	SiC-0	SiC-3.4	SiC-5.6	SiC-9.0	SiC-14.6
3 μm	0.284	0.826	0.788	0.889	0.916
10 μm	0.284	0.657	0.745	0.574	0.703
40 μm	0.284	0.582	0.619	0.594	0.539

Combinations of abrasives in different particle sizes probably result in an improved wear resistance for the friction composites. We observed that the particle sizes of abrasives affect the specific wear rate of friction composites. The model composites are prepared by adding combinations of Al_2O_3 particles in both big (μm)/big (μm) and big (μm)/small (nm) particle sizes [33]. The samples with abrasives in the big/small combination mode put up a more excellent performance in wear resistance, reflected as a lower specific wear rate, when compared with the other kind of composites.

EFFECTS OF ABRASIVES ON DISC LIFE

Misuse and/or excessive addition of abrasives in friction composites may result in a severe erosive damage to the disc due to the abrasive wear mechanism [34]. Discs used for brake systems are generally made of cast iron, whose Mohs hardness is lower than that of abrasives in brake linings.

Therefore, the iron element (Fe), especially in forms of strips or large chips, can be easily observed on the surface of non-metallic friction composites, which provides positive proof for destroying the brake disc with an abrasive effect of abrasives in the composites (Figure 3). Transferring iron from the disc to the surface of a friction composite can be considered as one of the main characteristics to understand the abrasive wear mechanism in a friction process between a disc-pad friction couple.

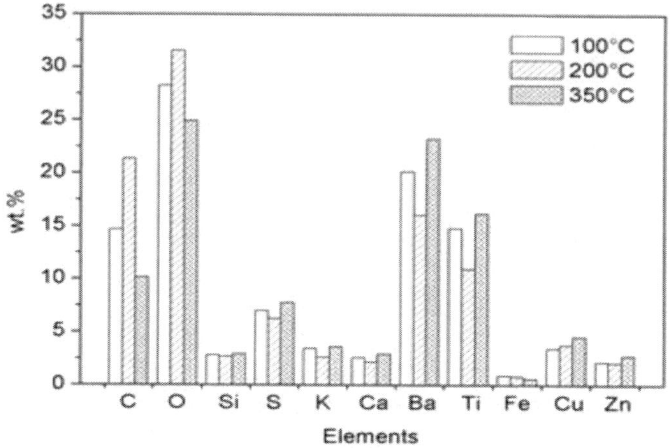

(a)

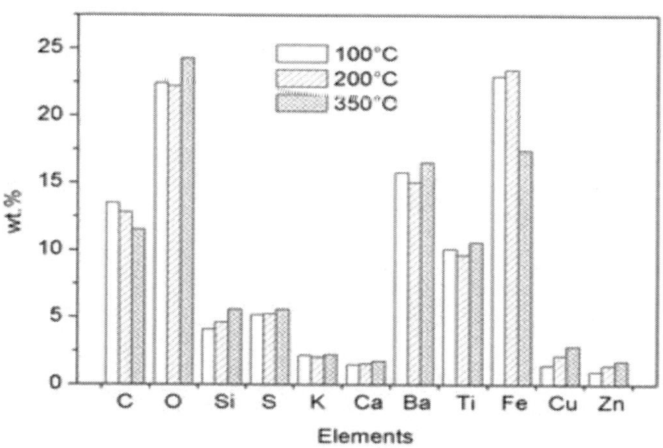

(b)

Figure 4. Continued on next page.

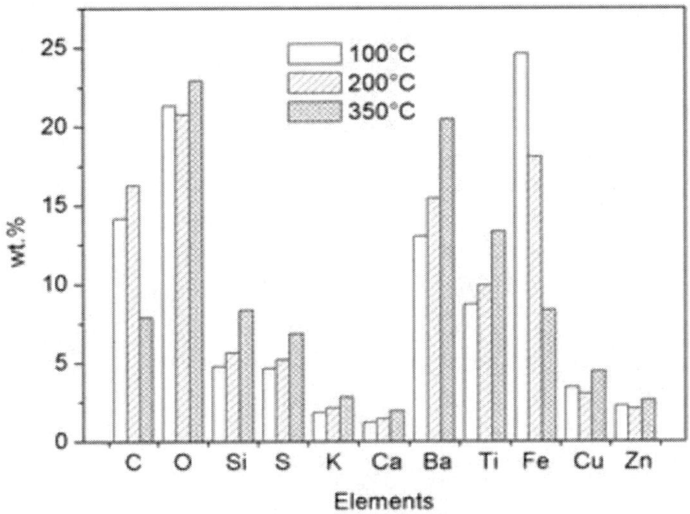

(c)

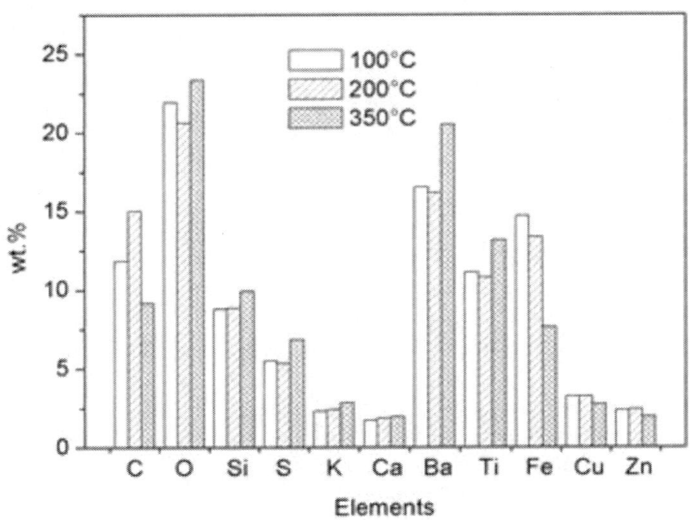

(d)

Figure 4. Effects of Particle Sizes of SiC Abrasive (a) SiC-0, (b) SiC-3.4 (F360-40 μm), (c) SiC-3.4 (F600-10 μm) and (d) SiC-3.4 (F1200-3 μm) on the Transfer of Iron.

The experiment result in Figure 4 shows consistence with the abrasion friction and wear model. Without addition of SiC abrasive, the amount of the iron element (Fe) transferred from disc to friction surface of the composites is significantly lower than that of samples with SiC particles. Besides, on the basis of adding the same amount of abrasive, the weight content of transferred iron also rose along with the particle size of SiC increased.

On the friction surface of the composites, the iron and its oxides have been observed mainly in two forms, the compacted iron debris (Figure 5a) and the iron stripes (Figure 5b). The compacted iron debris can be generated by iron-containing raw materials in bulk or transferred from the disc made of cast iron, while the iron stripes are almost cut from the disc by abrasives or ceramic fibers.

(a) (b)

Figure 5. Iron Debris around the Edge of Abrasives (a) and Iron Stripes (b) Observed on the Surface of Friction Composites.

EFFECTS OF ABRASIVES ON FRICTION LAYERS

Primary and Secondary Contact Plateaus

Friction mechanisms, related to the brake friction composites pressed against a rotating disc, are of great importance in understanding the relationships between composition and friction performance. The characterization of friction layers can be developed as effective means of investigating the friction mechanisms. As shown in Figure 5a and Figure 6, friction layers may be gradually formed parallel to the friction direction, when the wear debris can be continuously agglomerated and densified around the edge of an abrasive particulate, a ceramic fiber, or a steel wool. Obviously, the abrasives have an important influence on formation of friction layers. Concretely speaking, the exposed surface of the abrasive makes a contribution to the formation of a primary contact plateau, while its edge plays an important role in the formation of a secondary contact plateau [35]. Similar viewpoints can be found in the literature, in which the formation of the secondary contact plateaus by the compaction of wear debris was provided quite a reasonable description by Osterle et al. [36] or Muller et al. [37].

Figure 6. Formation of Hard Contact Patches: around the Edge of Ceramic Fiber.

Calculation of the Area of Hard Contact Patches

In order to calculate the area of the primary and secondary contact plateaus related to an abrasive particulate in the composite surface as shown in Figure 6, an elliptical model is introduced for simplicity (Figure 7). Assuming that the abrasive particulate is exposed on the composite surface with a circular cross section whose radius is assigned as b for the following computing. Therefore, the area of an abrasive (A_{abr}) can be roughly considered as the area of primary contact plateau, which is calculated as follows:

$$A_{abr} = \pi b^2 \tag{8}$$

The area of secondary contact plateau, formed by aggregation and compaction of the debris, can be assigned as A_{deb} and calculated by:

$$A_{deb} = (\pi ab - \pi b^2)/2 \quad (9)$$

Hence, the total contact area (A_p) is the sum of the above two parts, A_{abr} and A_{deb}:

$$A_p = A_{abr} + A_{deb} = \pi b^2 + \pi ab/2 - \pi b^2/2 = \pi b(a+b)/2 \quad (10)$$

Also, A_p can be calculated in another way:

$$A_p = \pi b^2/2 + \pi ab/2 = \pi b(a+b)/2 \quad (11)$$

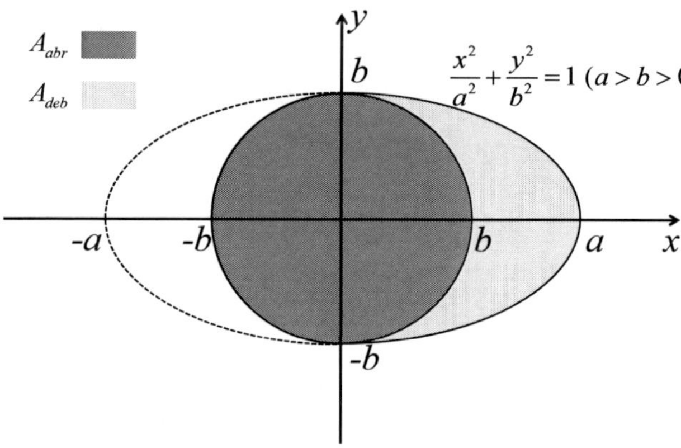

Figure 7. Elliptical Model for Calculation of the Area of Friction Layers Caused by Abrasives.

Simulation of Friction Layers Formation by Cellular Automata Model

For the purpose of simulating the formation of friction layers, the Cellular automata (CA) approach has been established by Ostermeyer et al. [37, 38] and Osterle et al. [39, 40]. The motion mode of the wear debris on the friction layers of a friction composite can be established by applying the cellular automata (CA) method [41, 42]. With a two-dimensional (2D) CA model, the movement of debris and the morphology of the formed friction layers were simulated, assuming that there is no wear for brake pads. When using the three-dimensional (3D) CA model, the corresponding simulation was developed under a wear condition, as shown in Figure 8, where the formation of new friction layers on new friction surface is shown due to wear. The formation mechanism of friction layers was caused by the debris orderly accumulated around the edges of abrasives along the friction direction. The results show that the CA method is not affected by the complex stress analysis and chemical reactions on the friction layer. When the rules and cellular neighbors are set well, the CA model can be used to provide reasonable simulations on the movement of the wear debris. Therefore, it demonstrates that the CA model could be widely utilized in the field of friction layer research, as well as it plays an instructive role in understanding the formation of friction layers.

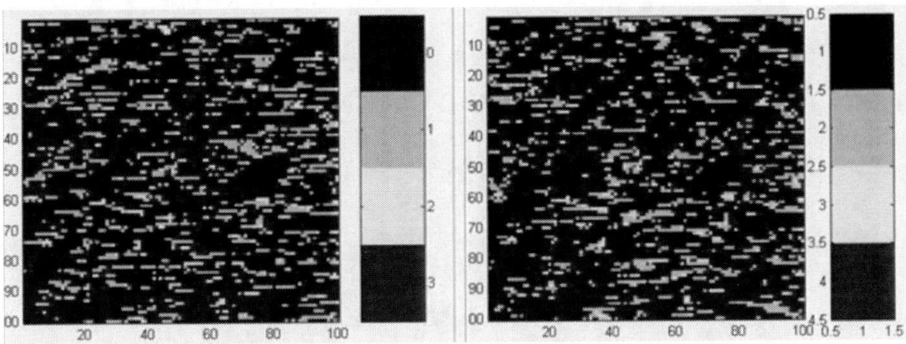

(a) (b)

Figure 8. 3D Simulation of Friction Layers Formation by CA Method: (a) Surface Layer and (b) 5th Layer.

CONCLUSION

Although the complex and non-linear friction performance related to abrasives in brake friction materials cannot be well described by using any single parameter such as hardness, particle size, and friction conditions (temperature, humidity, speed and load), the characteristics of the abrasion effect of the abrasives, both positive and negative roles, can be experimentally and theoretically investigated to some extent. The conclusions for the roles of abrasives in this study can be summarized as follows:

(1) Hardness, particle size and content of abrasives have positive influences on the friction coefficient of brake composites, i.e., the friction coefficient of the friction materials can be enhanced by adding

abrasives with a higher hardness, a larger particle size or a higher content of abrasives;

(2) The abrasive with a higher hardness could increase the wear of friction composites;

(3) Misuse and/or excessive addition of abrasives in friction composites may result in a severe erosive damage to the disc due to the abrasive wear mechanism, partly because abrasives may cause transferring iron debris/stripes from the brake disc to the surface of friction composites;

(4) Abrasives generally play an important role in forming hard contact patches, whose area can be calculated by introducing an elliptical model. Meanwhile, the cellular automata method has been proved to be an effective tool for simulating the forming mechanism of hard contact patches during a brake process;

(5) The critical volume fraction of abrasives properly exists and falls in the interval 0.034-0.056, which can be used to optimize formulations of friction materials in the future.

ACKNOWLEDGMENTS

Financial supports from the National Natural Science Foundation of China (50373002 and 50673012), the bilateral projects between China and Czech Republic (37-20, 38-7 and 39-9), and the Programs of International Cooperation granted by the Ministry of Education, Youth and Sports of Czech Republic (ME10121) are gratefully acknowledged.

REFERENCES

[1] Bijwe, J. (1997). Composites as friction materials: recent development in non-asbestos fiber reinforced friction materials – a review. Polymer Composites, 18, 378-396.

[2] Chan, D., Stachowiak, G. W. (2004). Review of automotive brake friction materials. Proceedings of Institution of Mechanical Engineering Part D: Journal of Automobile Engineering, 218, 953-966.

[3] Lu, Y., Tang, C. F., Wright, M. A. (2002). Optimization of a commercial brake pad formulation. Journal of Applied Polymer Science, 84, 2498-2504.

[4] Tang, C. F., Lu, Y. (2003). Optimization of an industrial friction material formulation. Tribology Transactions, 46, 19-23.

[5] Tang, C. F., Lu, Y. (2004). Combinatorial screening of ingredients for steel wool based semi-metallic and aramid pulp based non-metallic friction materials. Journal of Reinforced Plastics and Composites, 23, 51-63.

[6] Han, L. Huang, L, Zhang, J., Lu, Y. (2006). Optimization of ceramic friction materials. Composites Science and Technology, 66, 2895-2906.

[7] Lu, Y. (2006). A combinatorial approach for automotive friction materials: Effects of ingredients on friction performance. Composites Science and Technology, 66, 591-598.

[8] Lu, Y. (2002). A combinatorial approach for automotive friction materials: Combined effects of ingredients on friction performance. Polymer Composites, 23, 814-822.

[9] Lu, Y., Liu, Y., Fan, Y., Tang, C. F. (2010). Combinatorial friction materials research. In Bulk Materials: Research, Technology and Applications, Nova Publishers, 245-266.

[10] Lu, Y. (2003). A Golden Section approach to optimization of automotive friction materials. Journal of Materials Science, 38, 1081-1085.

[11] Lu, Y., Tang, C. F., Zhao, Y., Wright, M. A. (2004). Optimization of a semimetallic friction material formulation. Journal of Reinforced Plastics and Composites, 23, 1537-1545.

[12] Han, L., Liu, X., Lu, Y., Zhao, Y., Li, W., Zhang, Z. (2006). Combinatorial tribological materials research: five-component composite systems. International SAMPE Symposium and Exhibition Proceedings, on CD.

[13] Blau, P. J. (2001). The significance and use of the friction coefficient. Tribology International, 34, 585-591.

[14] Holmberg, K., Ronkainen, H., Laukkanen, A., Wallin, K. (2007). Friction and wear of coated surfaces – scales, modeling and simulation of tribomechanisms. Surface and Coatings Technology, 202, 1034-1049.

[15] Filip, P., Weiss, Z., Rafaja, D. (2002). On friction layer formation in polymer matrix composite materials for brake applications. Wear, 252, 189-198.

[16] Jang, H., Kim, S. J. (2000). The effects of antimony trisulfide (Sb_2S_3) and zirconium silicate ($ZrSiO_4$) in the automotive brake friction material

on friction characteristics. Wear, 239, 229-236.

[17] Lee, E. J., Hwang, H. J., Lee, W. G., Cho, K. H., Jang, H. (2010). Morphology and toughness of abrasive particles and their effects on the friction and wear of friction materials: a case study with zircon and quartz. Tribology Letters, 37, 637-644.

[18] Satapathy, B. K., Bijwe, J. (2004). Wear data analysis of friction materials to investigate the simultaneous influence of operating parameters and compositions. Wear, 256, 797-804.

[19] Han, Y., Tian, X., Yin, Y. (2008). Effects of ceramic fiber on the friction performance of automotive brake lining materials. Tribology Transactions, 51, 779-783.

[20] Lu, Y., Wang, K., Jiao, L., Matejka, V., Martynkova, G. S. (2008). Ceramic abrasives in brake friction materials. SAE 26th Annual Brake Colloquium and Exhibition Proceedings, 53-57.

[21] Ma, Y., Martynkova, G. S., Valaskova, M., Matejka, V., Lu, Y. (2008). Effects of $ZrSiO_4$ in non-metallic brake friction materials on friction performance. Tribology International, 41, 166-174.

[22] Matejka, V., Lu, Y., Fan, Y., Kratosova, G. (2008). Effects of silicon carbide in semi-metallic brake materials on friction performance and friction layer formation. Wear, 265, 1121-1128.

[23] Matejka, V., Martynkova, G. S., Ma, Y., Lu, Y. (2009). Semi-metallic brake friction materials containing $ZrSiO_4$: friction performance and friction layers evaluation. Journal of Composite Materials, 43, 1421-1434.

[24] Zhao, Y., Lu, Y., Wright, M. A. (2006). Sensitivity series and friction surface analysis of non-metallic friction materials. Materials & Design, 27, 833-838.

[25] Yun, R., Martynkova, G. S., Lu, Y. (2011). Performance and evaluation of nonasbestos organic brake friction composites with SiC particles as an abrasive. Journal of Composite Materials, 45, 1585-1593.

[26] Yun, R., Filip, P., Lu, Y. (2010). Performance and evaluation of eco-friendly brake friction materials. Tribology International, 43, 2010-2019.

[27] Fan, Y., Matejka, V., Kratosova, G., Lu, Y. (2008). Role of Al_2O_3 in semi-metallic friction materials and its effects on friction and wear performance. Tribology Transactions, 51, 771-778.

[28] Cho, K. H., Jang, H., Hong, Y. S., Kim, S. J., Basch, R. H., Fash, J. W. (2008). The size effect of zircon particles on the friction characteristics of brake lining materials. Wear, 264, 291-297.

[29] Bijwe, J., Aranganathan, N., Sharma, S., Dureja, N., Kumar, R. (2012).

Nano-abrasives in brake friction materials – influence on tribological properties. Wear, 296, 693-701.

[30] Vlastimil, M., Lu, Y., Jiao, L., Huang, L., Martynkova, G. S., Tomasek, V. (2010). Effects of silicon carbide particle sizes on friction-wear properties of friction composites designed for car brake lining applications. Tribology International, 43, 144-151.

[31] Tomasek, V., Kratosova, G., Yun, R., Fan, Y., Lu, Y. (2009). Effects of alumina in non-metallic brake friction materials on friction performance. Journal of Materials Science, 44, 266-273.

[32] Oleksowicz, S., Mruk, A. (2011). Friction process in the presence of hard abrasive particles – cooperation model. Surface Review Letters, 18, 249-259.

[33] Tang, C. F., Lu, Y. (2003). Effects of alumina particle sizes on wear. Friction and Sealing Materials, (4), 28-37.

[34] Talib, R. J., Muchtar, A., Azhari, C. H. (2003). Microstructural characteristics on the surface and subsurface of semimetallic friction materials during braking process. Journal of Materials Processing Technology, 140, 694-699.

[35] Eriksson, M., Jacobson, S. (2000). Tribological surfaces of organic brake pads. Tribology International, 33, 817-827.

[36] Osterle, W., Griepentrog, M., Gross, T., Urban, I. (2001). Chemical and microstructural changes induced by friction and wear of brakes. Wear, 251, 1469-1476.

[37] Muller, M., Ostermeyer, G. P. (2007). A cellular automaton model to describe the three-dimensional friction and wear mechanism of brake systems. Wear, 263, 1175-1188.

[38] Ostermeyer, G. P., Muller, M. (2006). Dynamic interaction of friction and surface topography in brake systems. Tribology International, 39, 370-380.

[39] Osterle, W., Klos, H., Urban, I., Dmitriev, A. I. (2007). Towards a better understanding of brake friction materials. Wear, 263, 1189-1201.

[40] Dmitriev, A. I., Osterle, W., Klos, H. (2008). Numerical simulation of mechanically mixed layer formation at local contacts of an automotive brake system. Tribology Transactions, 51, 810-816.

[41] Xu, Y., Lu, Y., Huang, J. (2010). Simulation of friction layer formation by cellular automata model. Journal of Beijing University of Chemical Technology, 37(6), 136-140 (in Chinese).

[42] Fan, Y., Lu, Y. (2010). Analysis of wear debris during friction. Lubrication Engineering, 35(12), 17-21 (in Chinese).

In: Recent Advances in Abrasives Research
Editor: Dirk Bähre
ISBN: 978-1-62808-566-2

Chapter 3

MULTIPLE INTERACTIONS: A NEW APPROACH TO SIMULATE ABRASIVE WEAR

W. M. da Silva, H. L. Costa and J. D. B. de Mello*
Laboratório de Tribologia e Materiais, Faculdade de Engenharia Mecânica, Universidade Federal de Uberlândia, Uberlândia-MG, Brasil

ABSTRACT

The mechanisms that result in material removal by abrasive particles depend on the particle dynamics at the interface. The rolling of abrasive particles between the surfaces is evidenced by the presence of indentations on the worn surfaces, and the sliding of active particles on the sample surface produces scratching and/or ploughing. Moreover, for both particle dynamics (rolling or sliding), microcracking can occur when the contact stresses achieve a threshold limit. The particle dynamics can vary with the imposed tribological conditions. In this chapter, we present a new methodology for the simulation of abrasive wear that involves well-controlled instrumented indentation and/or scratch tests, where indentations represent the contact of rolling abrasive particles without sliding, which can lead to deformation or cracking, and scratches represent the sliding of abrasive particles, which can lead to ploughing, edge formation, cutting, or cracking. The equipment especially developed to reproduce the action of abrasives particles consists of two horizontal

* Corresponding author: J. D. B. de Mello. Laboratório de Tribologia e Materiais, Faculdade de Engenharia Mecânica, Universidade Federal de Uberlândia, Av. João Naves de Ávila 2121, Campus Santa Mônica, Uberlândia-MG, 38400-901, Brasil. E-mail: ltm-demello@ufu.br.

sliders (0.1 μm resolution) that drive the sample, a piezoelectric translator with a resolution of 5 nm to control the indenter and a 3D load cell to control and measure the load. Initially, the surface topography of a real worn surface was assessed by laser interferometry and SEM analysis was used to determine its prevailing wear mechanisms, which were then used as reference for the simulation tests. The simulation methodology here proposed produced a surface topographically and morphologically similar to that of the reference. The results showed that the superimposition of indentations had a great effect on the morphology and topography of the simulated surfaces. In this sense, a new parameter was introduced to evaluate wear mechanism transitions, which was called superimposition of interactions. In order to study the effect of superimposition, series of indentations and parallel scratches were created by varying the distance between mono-interactions and the size of the mono-interactions. Different materials with very distinct behaviour were chosen in order to scan a wide range of possible transitions: ductile annealed copper, tool steel and brittle soda lime glass.

Keywords: Wear mechanisms, indentation, scratches, simulation

1. INTRODUCTION

Abrasive wear is a very complex interfacial process that can evolve rapidly and therefore it is mainly studied using the experimental approach. The interactions between the hard particles/protuberances and the surface(s) involved are highly complicated due to the irregular shapes and sizes of the abrasive agent, generating complex surface topographies on the worn surfaces.

Abrasive wear is defined as material loss caused by the relative motion of surfaces which have: hard protuberances in either of the moving surfaces, or the presence of free abrasive particles between them, or fixed abrasives particles in either surface [1]. Abrasive wear was primarily divided into two main modes: two-body and three-body abrasive wear [2]. More recently, abrasive wear mode has been classified according to the abrasive particle dynamics at the interface [3-5]:

(i) Sliding of active abrasive particles at the interface;
(ii) Rolling of abrasives between the surfaces.

The sliding of fixed particles or hard protuberances causes scratches on the antagonist surface.

A sequence of scratches may be used to simulate the sliding of the particles. According to classic literature [6-8], each scratch induces the mechanisms of microcutting and/or microploughing and/or microcracking.

The prevailing wear mechanism has been shown to be related to the movement of the active particles present in the interface [5-7]. The particle dynamics (sliding or rolling) can vary according to the tribological conditions in either laboratory tests or real applications. Adachi and Hutchings mapped abrasive particle dynamics as a function of tribological parameters in micro abrasion tests [1].

Gore and Gates [9] studied the movement of abrasive particles in different abrasive wear tests, correlating it to the prevailing wear mechanisms. These authors suggested side-fin and wedge formation as intermediate wear mechanisms between microploughing and microcutting. They argued that "pure" microcutting rarely occurs in practical wear situations.

The rolling of abrasive particles in the interface leads to multiple indentations. The rolling of abrasive particles can be simulated using a sequence of indentations. Each indentation may induce ductile or brittle behaviour. Rolling and sliding together cause a mixture of indentations and scratches on the surfaces.

Kayaba et al. [10] carried out scratch tests using a hard spherical pin inside a scanning electron microscope and the scratching process was observed successively. Their observations allowed them to define four types of abrasive wear for ductile materials: *i-* cutting-type wear, with the formation of ribbon-like wear debris, very similar to that of machining with a tool; *ii-* flaking-type wear, where many thin flakes pile up at the groove end and are detached as loose wear debris; *iii-* wedge formation, where a wedge grows ahead of the pin. Only the initial wedge forms wear debris, but once it is formed it remains there and no new wear debris is formed; and *iv-* ploughing wear, where wear is negligible. They modelled the transition between these wear types with the degree of penetration.

Williams and Hyncica [11, 12] correlated abrasive wear with the movement of the abrasive particle at the interface in lubricated contacts. They proposed that the particle movement defines the material removal mechanism, which in turn depends on the particle size and shape and on the distance between the sliding surfaces, thus promoting wear transitions from rolling to ploughing to cutting. Buijs et al. [13, 14] investigated abrasive wear mechanisms during the lapping of optical glasses. They proposed a model where the rolling of abrasive particles can lead to material removal when the load per particle achieves the threshold value for lateral cracking.

They also showed that the particle dynamics depends on the counter-body hardness, since for softer counter-bodies, particle sliding becomes more significant than rolling.

Fang et al. [15, 16] registered the movement of a single particle when inserted between two moving surfaces. They showed that the shape of the abrasive particle also determines the particle dynamics: round particles are more prone to sliding, and angular particles tend to roll between the surfaces.

The hardness of the particle is connected to its capacity of fragmentation. A harder surface causes particle degradation [8, 17]. When paired surfaces present high hardness differences, active particles tend to embed on the softer surface. This fact promotes the sliding of the particle on the harder surface [11, 12].

The most common factors that interfere on the active particles dynamics can be summarized in the scheme presented in Figure 1.

Sclerometric tests use single particles to simulate abrasive wear, and therefore allow good control of particle dynamics. Williams and Xie [18] investigated the occurrence of wear resulting from the interaction between parallel scratches. Initially, they carried out isolated scratches and showed a transition between ploughing and cutting by changing the attack angle. Reducing the distance between scratches, the wear mechanism changed in some cases, and this was attributed to the interaction between neighbouring scratches.

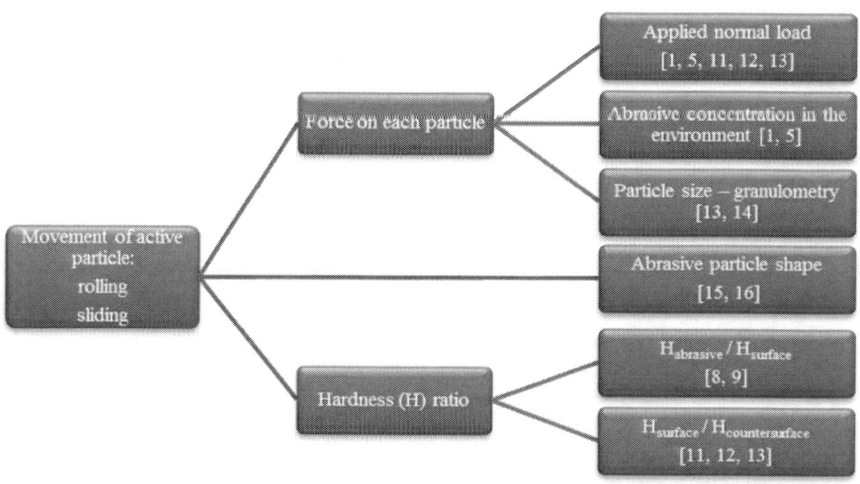

Figure 1. Tribological parameters affecting the particle movement at the wear interface.

However, Williams [19] points out later that studies involving the interaction between successive wear events or their statistical combination to generate observed final topographies have been scarce in the literature.Mezlini et al. [20] used the interaction between single scratch tests to detect transitions in wear mechanisms for Aluminium alloys.

In their experiments, they produced two parallel scratches with a defined distance between them and then a third parallel scratch in the middle of the previous ones. They found that the wear mechanism changed from ploughing and wedge formation for a single scratch to cutting when the scratches interacted.

A new methodology for representing the movement of abrasive particles using a sequence of well controlled interactions has been developed by da Silva and de Mello [21]. This new method is based on the reproduction of the particle dynamics. An indentation represents the contact of a rolling abrasive particle, which can lead to deformation or cracking, and a scratch represents the sliding of an abrasive particle, which can lead to ploughing, side-fin, wedge formation, cutting or cracking.

More recently, da Silva and de Mello [22] reported that a real worn surface obtained in a rubber wheel test could be fairly well reproduced using a sequence of parallel scratches. The real worn surface and the surface simulated using parallel scratches were morphologically and topographically similar. Initially, a special routine was created to ensure random positioning of the interactions and a minimum coverage of the indented area. Since the positioning of the indentations was random, it was possible to control the normal load and the number of indentations, but neither the degree of superimposition nor the size of the interactions. Applying such a routine to AISI 1010 steel samples, simulated surfaces could be produced, which had a roughness (S_q) equivalent to the reference surface obtained in a rubber wheel test, but a different morphology. In fact, the simulated surfaces were quite uniform, with small indentations of equivalent size, whereas the reference surface was constituted of multi indentations with different sizes. This fact induced the use of variable normal loads as a possible solution, and a specially developed routine, which could generate a randomly distributed profile of loads, was developed.

The application of this load profile to the simulation produced surfaces that were morphologically more similar to the reference, but presented a different roughness. Similar results were obtained for parallel scratches. As a consequence, a new parameter was introduced and it was called superimposition of interactions [22, 23].

The experimental simulation method here proposed has the advantage of reproducing real worn surfaces where information about all the forces associated to the creation of each individual event that composes the final worn surface is stored. It could be used simply to rank materials for a specific tribosystem, but it also opens new investigation routes that could lead to a more fundamental understanding of abrasive wear. In this chapter, this new approach is used to investigate transitions in abrasive wear mechanisms as a function of the superimposition of interactions. In order to study the effect of superimposition, series of indentations and parallel scratches were created by varying the distance between mono-events. On the other hand, the experimental simulation here proposed allows to envisage a new methodology for realistic computational simulation of abrasive wear, due to the knowledge of the precise location of each event, of the forces involved and of the depth resulting of the interaction of the abrasive particle with the sample surface.

2. SIMULATION

Figure 2 summarizes the new approach to simulate abrasive wear using sequences of scratches and/or indentations on a sample surface.

Initially, a worn sample obtained from either a laboratorial abrasive wear test or a field tribological situation, is used as the reference for the simulation. The assessment of the reference surface uses SEM to identify the occurrence of indentations and/or grooves/scratches, i.e., the prevalence of pure sliding, pure rolling, or a mixture of rolling and sliding. Moreover, 3D surface topographical parameters quantify the surface topography of the worn surface. The size of this sample area depends on the abrasion severity, since it needs to be large enough to characterize well the prevailing wear mechanisms. Laboratorial wear tests that uses small abrasive particles lead to relatively small events (indentations and/or scratches) and therefore areas of 1 mm^2 are large enough to characterize the wear mechanisms.

The prevailing mechanisms found in the reference surface will define the configuration of the simulation to be used. Rolling of the abrasive particles is simulated using sequences of indentations, whereas the sliding of abrasive particles is simulated using sequences of scratches. If a mixture of rolling and sliding is found in the reference surface, the simulation should involve sequences of both scratches and indentations. During the simulation tests, the position and direction (for the case of scratches) of each mono-event can be precisely controlled.

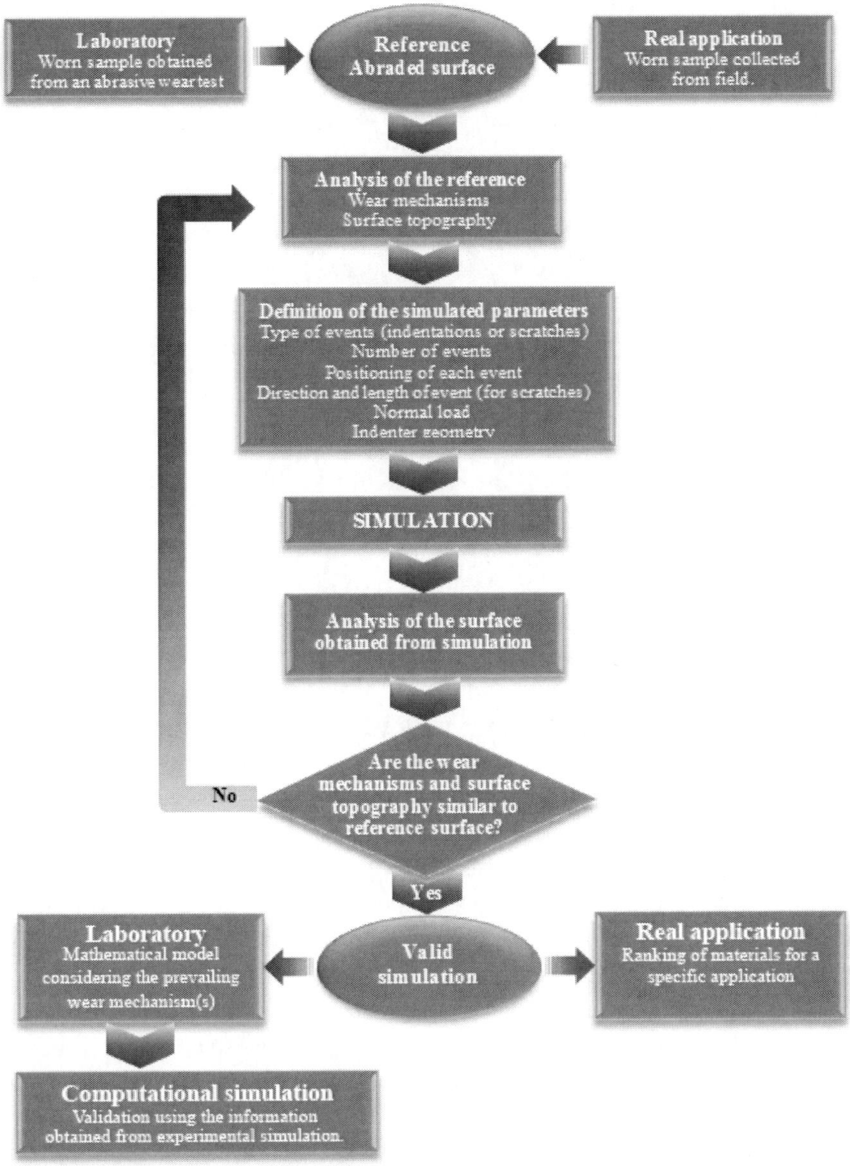

Figure 2. Flowchart of the abrasive wear simulation.

The quantitative analysis of the surface topography of the reference provides information about each event to be created during the simulation. Surface topography parameters define the depth and therefore the normal load to be used to create each event. The sequence of superimposed events alters the sample surface in order to reproduce the wear mechanism found in the reference. The simulation is considered valid if both the topography and the morphology of the surface created during the simulation are similar to those for the reference; otherwise, the simulation parameters must be redefined. The similarity between the simulated surface and the worn surface is assessed by surface topography quantitative parameters such as the root mean square roughness (Sq) and the root mean square wavelength (λq), by SEM identification of the particle dynamics (rolling or sliding), by the SEM identification of the wear mechanisms (microploughing, wedge formation, microcutting or microcracking), and by SEM visualization of the surface morphology.

The main input information is the surface topography of the reference. The surface roughness parameters define the simulation parameters. If the simulation is valid, the main output information is obtained: wear rate (volumetric or gravimetric) and surface topography parameters. This information can be used, for example, to rank materials for a specific tribological situation.

Moreover, the simulation stores information about all the forces associated to each individual event. The knowledge of the precise location of each event, of the forces involved and of the depth resulting from the interaction of the abrasive particle with the sample surface opens a new route for the development of a realistic computational simulation of abrasive wear.

A micro-simulator was developed to reproduce the interaction of abrasive particles with a surface and is shown in Figure 3.

The device, Figure 3, operates with two high resolution (0.1 μm) sliders (slider x, slider y) that drive the sample horizontally while a Vickers indenter is moved vertically by another slider (slider z), also with a resolution of 0.1 μm.

Besides this, a piezoelectric translator (PZT) is used to control the indenter movement with a resolution of 5 nm for a maximum travel distance of 40 μm. A 3D load cell controls the intensity of the normal load and measures the applied loads and moments in the directions x, y and z. The resolution of the load cell is 0.001 N and the maximum normal load is 18 N. A specific developed Labview® interface controls the movements of sample and indenter. Further information about this device can be found in [22].

Figure 4a shows a configuration where the indentation position (*x, y*) is randomly defined inside the simulation area. It represents pure rolling of the particles, which leads to multiple indentations.

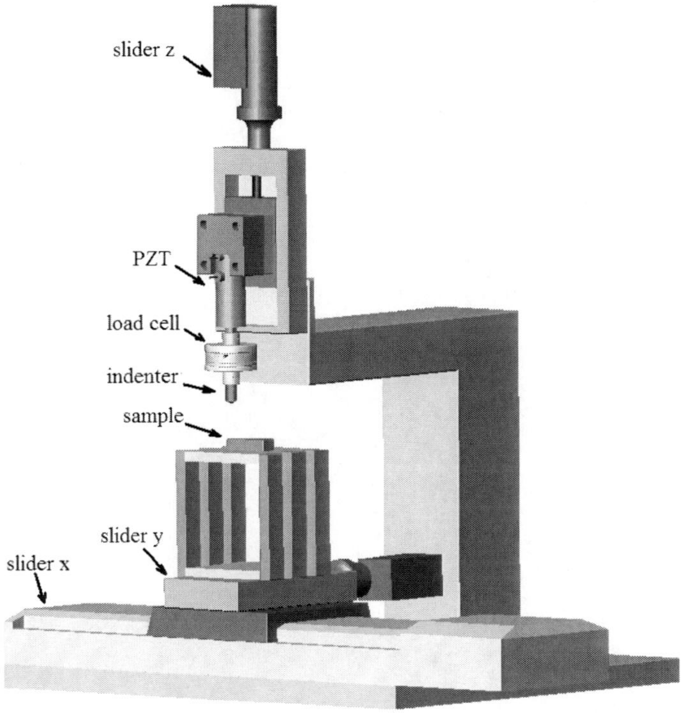

Figure 3. Scheme of the micro-simulator.

For random scratching, two randomly chosen points (*x, y*) define the initial and final position of sliding, Figure 4b. This configuration represents pure sliding of abrasive particles with no specific direction for the movement.

For a parallel scratching configuration, only the x-coordinate is randomly chosen, Figure 4c. It represents the sliding of abrasive particles with strong defined direction. Figure 4d shows a configuration involving both parallel scratching and indentations. It represents the simultaneous occurrence of parallel sliding and multiple indentations on the worn surface. Figure 4e shows a configuration with both random scratching and indentation and it represents the concomitant occurrence of rolling and sliding without specific direction of the particle movement. For the two last configurations (d and e), each time the simulation creates a single event, the program decides randomly if a scratch or an indentation will be used.

The first reference surface was obtained from a three-body lapping-like test on an AISI 1010 carbon steel. This test is schematically presented in Figure 5 [17, 24].

In this test, three samples are positioned in the sample holder, which allows a uniform distribution of the normal load (NL). The system is fed by a constant flux of dry abrasive.

The counterbody (disc) rotation induces the rotation of the sample holder. This test dynamics ensures the renewal of abrasive in the contact interface [17, 24].

Before each set of test the counterbody was conditioned by a one hour test using standard steel samples. This procedure guaranteed the same counterbody roughness before tests.

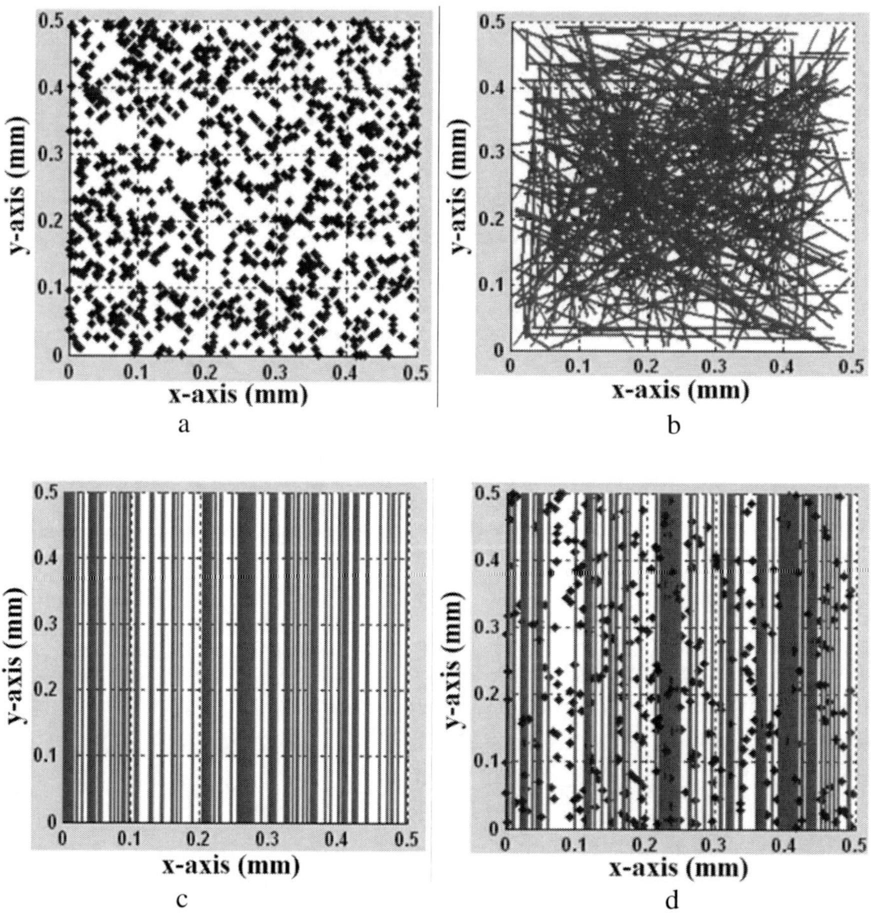

Figure 4. (Continued).

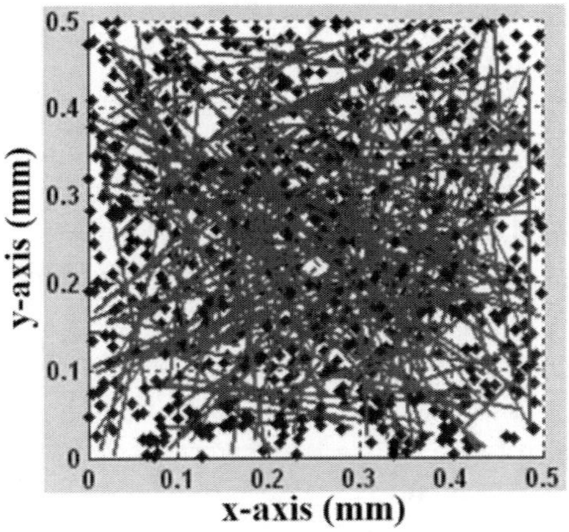

e

Figure 4. Schematic representation of possible test configurations.

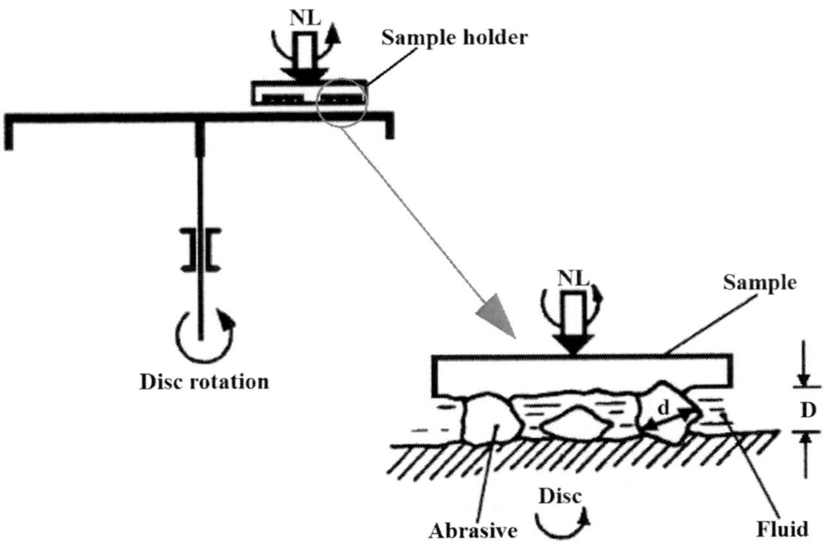

Figure 5. Schematic representation of a three-body lapping-like test [17, 24].

The tribological conditions were chosen in order to favour rolling of the abrasive particles. The normal load (NL) was 3.8 N, the counterbody was white cast iron (HV = 7365 MPa), and SiO_2 particles with a size range between 0.150 to 0.300 mm were used as abrasive particles. The samples surfaces were ground to Ra value of 0.5 μm and Rq value of 0.7 μm before the rubber wheel test. These parameters, after 30 minute test, resulted in a S_q value of 2.80 μm on the worn surface (Figure 6f).

In order to reproduce the reference surface, the configuration using random indentations was used (Figure 4a) and the simulation parameters were varied in order to obtain the same value of S_q found in the reference surface (Figure 6a to e). For this first simulation approach, AISI 1010 samples were ground and mirror polished before tests. The particle shape was assumed to be close to that of a Vickers indenter. This obviously leads to uniform tip geometry in the indentations and therefore does neither take into account possible breaking of the abrasive particles, which would change their geometry, nor the possibility of the presence of more rounded particles during abrasion. Future developments of the present work should use abrasive particles with different geometries. The normal indentation load (NL) and the amount of indentations were varied between tests. Despite reproducing the surface roughness of the reference surface, the size of the events found in the simulated surfaces was smaller.

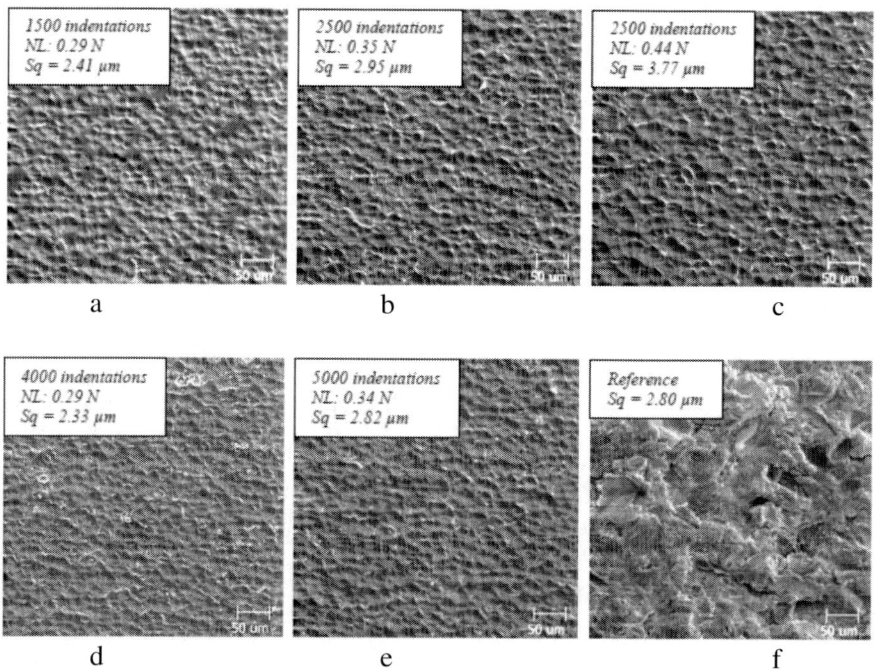

Figure 6. SEM images of simulated surfaces using random indentations on AISI 1010 carbon steel and comparison with the reference surface.

The reference surface shows events of different sizes. Such heterogeneity must be associated to the size distribution of the abrasive particles that act in the sliding surfaces. As shown in the model by Williams and Hincyca [11, 12], particle size determines the actual normal load per particle on the moving surfaces. The simulation, on the other hand, used a constant normal load.

A second reference surface was obtained from a rubber wheel test on M2 tool steel (HV = 6800 MPa). This device is schematically presented in Figure 7.

A normal load of 18.75 N was applied to the sample by a lever system. A continuous flow of dry SiO_2 particles (size: 0.3 to 0.6 mm) fed the system. The counter-body was a carbon steel disc covered with a rubber ring (styrene-butadiene - hardness 60 Shore A). The rotation of the counter body, fixed in 100 rpm, ensured fresh abrasives at the contact interface. The duration of each test was 30 minutes. The specimen surfaces were ground to Ra value of 0.5 μm and Rq value of 0.7 μm before the rubber wheel test. These test parameters favoured the sliding of abrasive particles, which led to parallel scratches distributed on the worn surface, as presented in Figure 8. This surface results from the action of particles with different sizes and shapes. In abrasion, the particle size distribution corresponds to the variation in depth of the observed scratches.

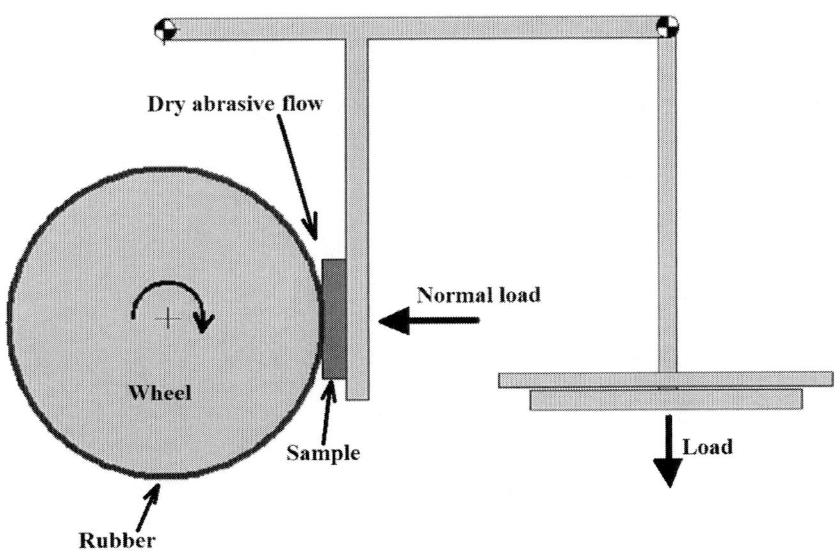

Figure 7. Scheme of the rubber wheel abrasive wear test.

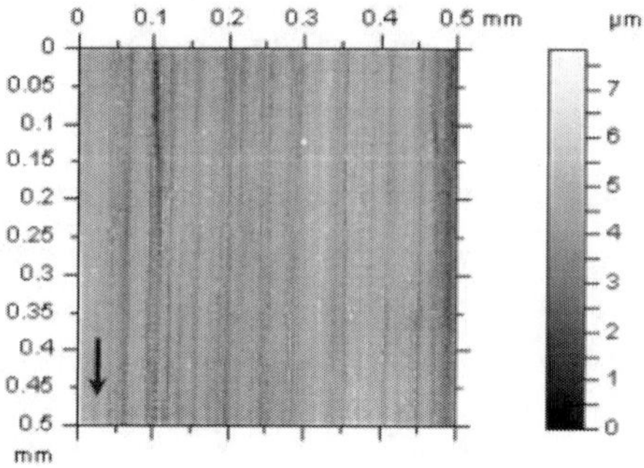

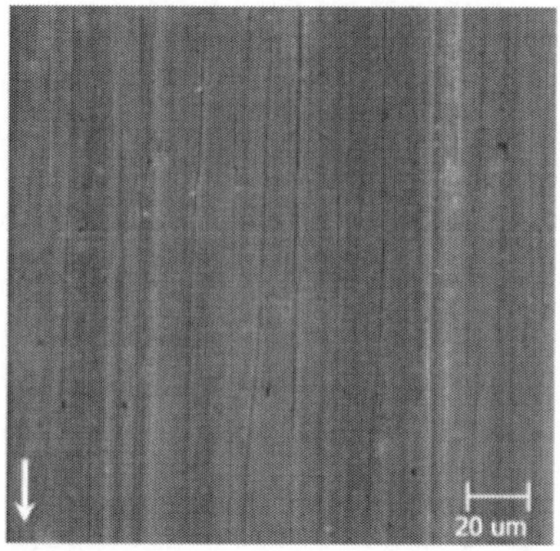

a b

Figure 8. Reference surface obtained for the rubber wheel test on M2 tool steel; (a) LI; (b) SEM.

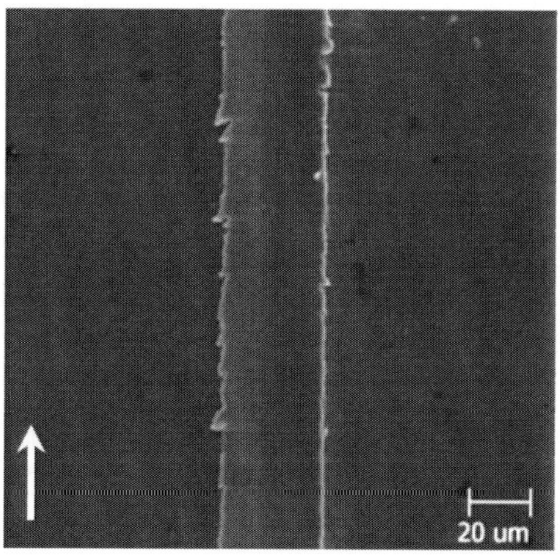

a

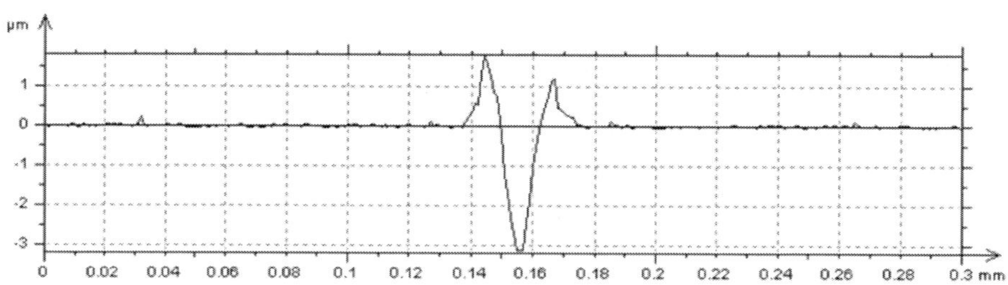

b

Figure 9. (a) Single scratch in tool steel, normal load 2.9 N, SEM; (b) cross section. LI.

A mean value of S_q = 0.83 μm was obtained from the laser interferometry (LI) measurements of the reference surface (Figure 8a).

Figure 9 shows an example of a single scratch carried out to establish a relationship between the normal load and the depth of the scratches, obtained from cross section measurements (Figure 9b).

The measurements of the single scratches for ten different normal loads allowed the calculation of the relationship between normal load and scratch depth for the used Vickers geometry, as shown in Figure 10. Using the transfer function shown in this figure, the mean depth of the scratches found in the reference surface (S_q = 0.83 μm) should be obtained using a normal load of 0.36 N.

Figure 11 shows a surface produced with a normal load of 0.36 N during the simulation. The value of S_q measured for this surface was 0.87 μm, which is fairly close to that found for the reference surface (0.8 μm, Figure 8a). However, the morphology of this surface is visually different from that of the reference surface.

This shows that for simulations where the depth of the scratches is obtained using a fixed normal load defined by the value of S_q of the reference, surface roughness is properly reproduced, but not surface morphology.

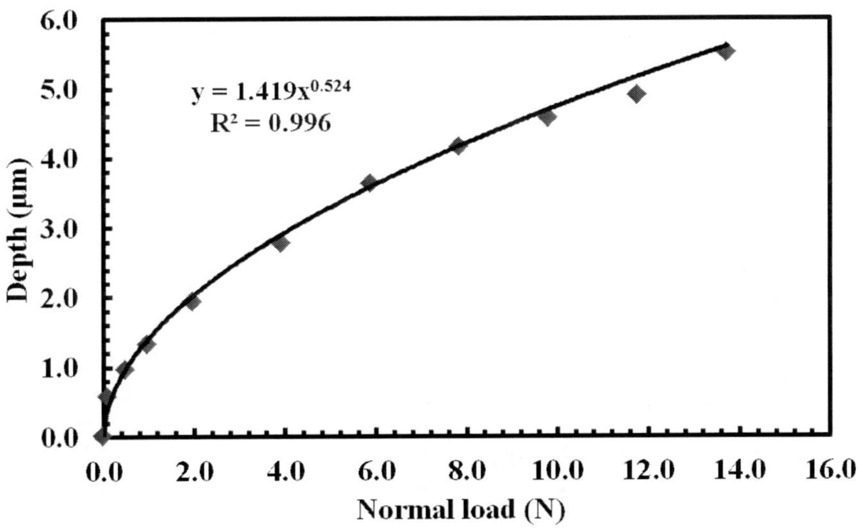

Figure 10. Influence of normal load on the depth of single scratches on M2 tool steel.

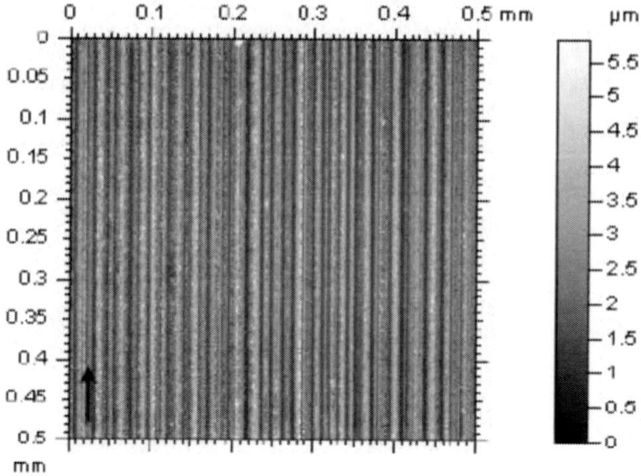

Figure 11. Simulation of parallel scratching using a constant normal load of 0.36 N; LI.

Two different aspects of the simulated surface need to be further analysed:

i The use of a constant load to generate the scratches produced a surface where all scratches have almost the same size, which is different from the topography observed in the reference surface, where the worn surface was generated by abrasives of different sizes and/or by different loads per particle. Clearly, each interaction size in the reference surface was caused by different interaction intensities for each abrasive particle. In the simulation process, the intensity of the interaction was controlled by the normal load. Therefore, in order to reproduce the scratch size variation observed in the reference surface, the intensity of the scratches in the simulation process needs to be variable; and

ii The root mean square roughness (Sq) of a surface represents an average variation in the height of irregularities [25], but it does not give information about the surface morphological aspect. This suggests that the reference surface should be better represented by a parameter that could capture the lateral distribution of the surface irregularities, such as the hybrid bidimensional parameter λq, which represents the root mean square wavelength of the profile [26, 27]. Considering that λq can be associated with the width of the scratches, a relationship was found between scratch width and normal load using single scratch experiments, as shown in Figure 12.

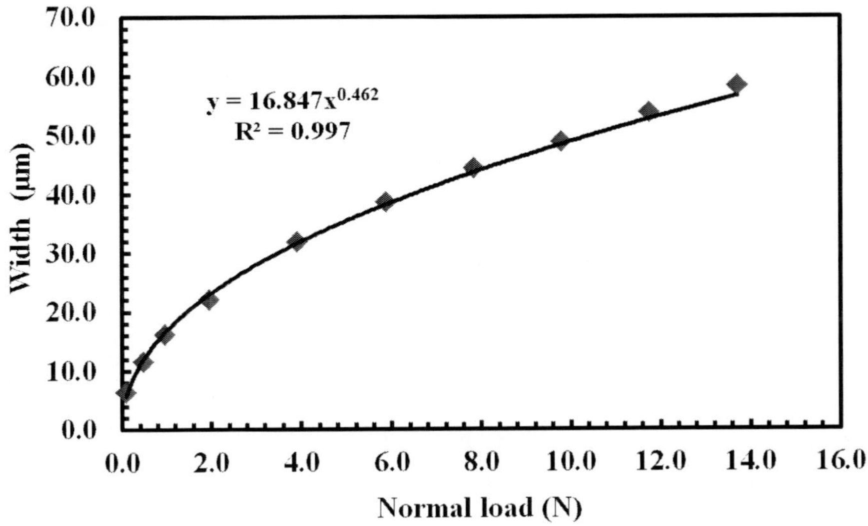

Figure 12. Effect of normal load on the width of single scratches on M2 tool steel.

Table 1. Equivalent normal load as a function of the root mean square wavelength (λ_q)

	λq (μm)	Equivalent normal load (N)
Minimum	10.8	0.38
Average	11.7	0.46
Maximum	17.5	1.08

The hybrid bidimensional parameter λq is calculated from the quadratic average amplitude (*Rq*) and from the quadratic average slope (Δq) as defined in Equation 1 [26, 27].

$$\lambda_q = 2\pi \frac{R_q}{\Delta_q} \qquad (1)$$

λ_q of all the profiles that compose the reference surface were calculated and are shown in Table 1. These values were used in a random generator to calculate the load distribution to be used in the tests, which is shown in Figure 13.

Figure 14 shows the simulated surface produced using the load distribution described in Figure 13, which looks very different from the reference surface.

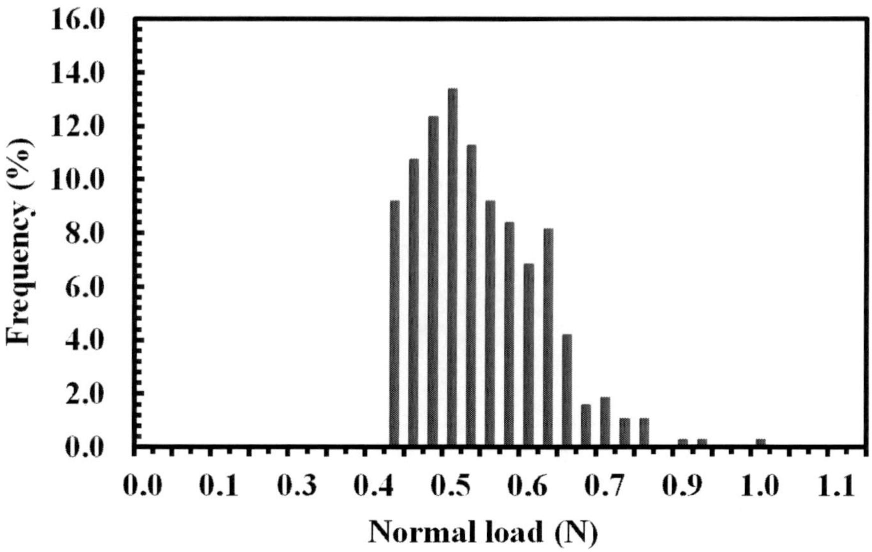

Figure 13. Load distribution to be used in the simulation tests with variable load.

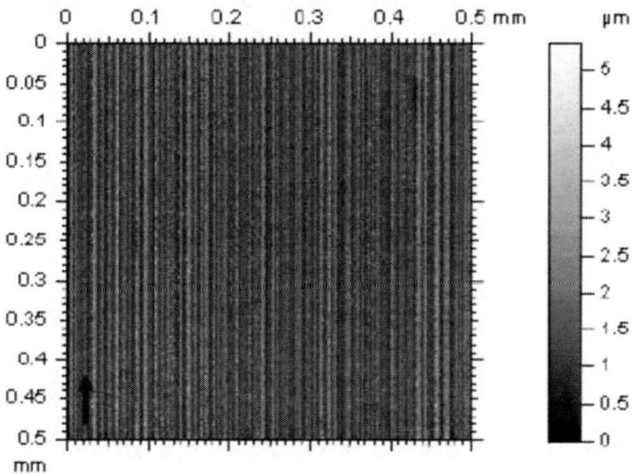

Figure 14. Parallel scratch simulation using variable normal load defined by the λ_q values; LI.

Indeed, the values for λ_q in the simulated surface and the reference surface were very different: 7.15 and 11.7, respectively.

This indicated that the use of a variable load and of λ_q as the controlling parameter is not sufficient to reproduce the surface morphology. In fact, it seems that the interaction between scratches affects the resulting topography.

2.1. Superimposition Concept

In order to study the interaction between individual events, the concept of superimposition of interactions was proposed. This concept applies to both indentations and scratches. For simplicity, initially the superimposition concept was applied to indentations, where sequences of indentation lines were carried out. To develop this concept, the first step was to conduct tests with individual indentations at different normal loads to establish the correlation between normal load and indentation width. Indentation width was measured using optical microscopy. In sequence, the actual indentation tests used a normal load of 2.30 N, which resulted in an indentation width (IW) of 18.35 μm. The length of the line of indentations was set as 300 μm. In order to vary the superimposition of events, the amount of indentations was increased progressively, decreasing the distance between indentations (DI), as shown in Figure 15a. The degree of superimposition (DS) was calculated according to Equation 2.

$$DS = \frac{(IW - DI)}{IW} \cdot 100 \tag{2}$$

To apply the concept of superimposition to parallel scratches, initial tests with individual scratches at different normal loads established the relationship between the normal load and the scratch width. The scratch width (SW) was measured by optical microscopy. By knowing the distance between scratches (D) and the scratch width, it was possible to control the degree of superimposition between the mono-events during the experiments, as shown in Figure 15b. Analogous to Equation 2, Equation 1 was used to calculate the degree of superimposition (DS) for parallel scratches.

$$DS = \frac{(SW - D)}{SW} \cdot 100 \tag{3}$$

Finally, the concept of superimposition was applied to indentations randomly distributed on a brittle surface, in order to identify how the proximity of stress fields of the indentations could lead to spalling with the increase of the number of indentations and of the normal load, and therefore of the superimposition of the events. Further details of these tests will be presented in Section 2.3.

2.2. Superimposed Parallel Indentations and Scratches in Ductile Materials

Initially, in order to verify the effect of the superimposition of parallel scratches and indentations, simulation tests were mainly carried out on M2 steel samples (HV = 6800 MPa). Sequences of both parallel scratches and indentations were used to simulate different abrasive wear mechanisms. The samples were mirror-polished before the simulations. In addition, some scratch tests for copper samples aimed to investigate further the superimposition effect for very ductile materials and therefore strain hardening effects. Before the simulations, the copper samples were annealed at 500 °C for 1h, which resulted in an average hardness of 471 MPa and then mirror polished.

Table 2 summarizes the conditions used in the simulations using lines of indentations on tool steel samples.

Table 2. Test conditions for the indentation tests on tool steel samples

Degree of superimposition %	NI	DI
-100	8	36.7
-50	11	27.5
-30	13	23.8
-10	15	20.2
0	16	18.3
10	18	16.5
20	20	14.7
30	23	12.8
40	27	11.0
50	33	9.2
60	41	7.3
70	54	5.5
80	82	3.8
90	163	1.8
98	817	0.4

NI = number of indentations; DI = distance between indentations (μm).

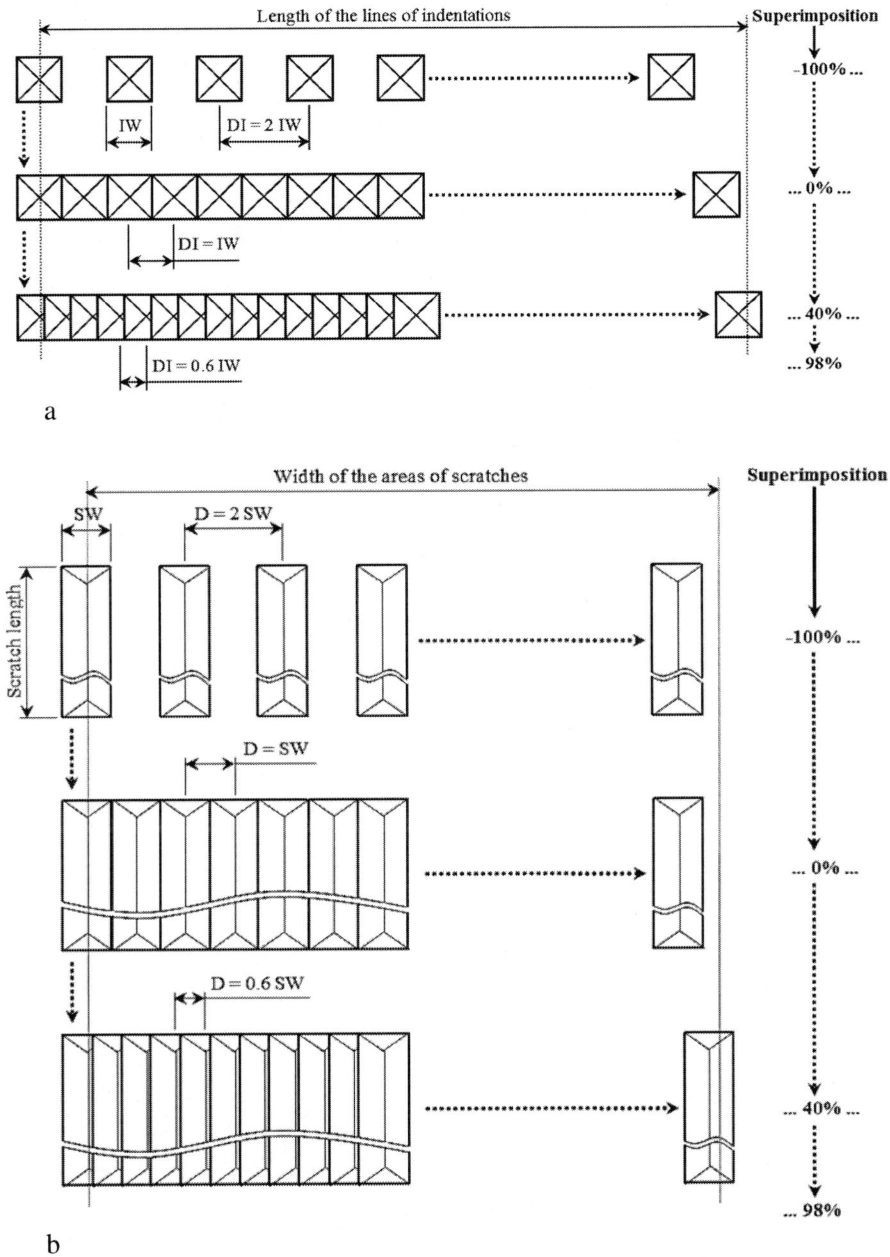

Figure 15. Schematic representation of the simulation tests (a) indentation lines with different number of indentations; (b) sequences of scratches.

Table 3 summarizes the conditions used for the simulation using parallel scratches on tool steel samples. In this table, a degree of superimposition of zero represents that the edges of each individual events are touching and therefore no superimposition occurs.

Table 4 summarizes the conditions for the additional scratch tests on copper samples, where negative degrees of superimposition mean that there is a separation between the individual events. An area of 500 μm x 1 mm (scratch length) was generated for each simulation.

After the simulations, cross sections of the tested samples were prepared. The very soft copper samples required a protection using hard nickel plating before mounting in a polymeric resin. An aqueous ammonia solution (5%) was used to etch the copper cross sections. Worn surfaces and wear mechanisms were analysed by optical micrography (OM), scanning electron microscopy (SEM) and laser interferometry (LI).

The correlation between normal load and indentation width for the tests with individual indentations at different normal loads is shown in Figure 16. This relationship allows the control of the indentation area, defined by the normal load imposed during the interaction. The value of indentation width (Figure 16) was applied in Equation 2 in order to control the superimposition degree during the superimposition tests.

Table 3. Sliding conditions for parallel scratching tests on tool steel samples

Normal load (N)	0.5		5.9		11.8	
Degree of superimposition %	N	D	N	D	N	D
0			13	38.3		
20			16	30.6		
40			22	22.0		
50	82	6.11	26	19.14	19	26.37
60	102	4.89	33	15.31	24	21.10
70	136	3.67	44	11.48	32	15.82
80	204	2.45	65	7.66	47	10.55
85	273	1.83	87	5.74	63	7.91
90	409	1.22	131	3.83	95	5.27
95	818	0.61	261	1.91	190	2.64
98	2045	0.24	653	0.77	474	1.05

N = number of parallel scratches; D = distance between scratches (μm).

Table 4. Sliding conditions for parallel scratching tests on copper samples

Normal load (N)	0.5		3.0		6.0	
Degree of superimposition %	N	D	N	D	N	D
-100			3	199.9		
-50			3	149.9		
0			5	99.9		
25			7	75.0		
50	29	17.37	10	49.98	7	75.23
60	36	13.89	13	39.98	8	60.18
70	48	10.42	17	29.99	11	45.14
80	72	6.95	25	19.99	17	30.09
85	96	5.21	33	14.99	22	22.57
90	144	3.47	50	9.98	33	15.05
95	288	1.74	100	4.98	66	7.52
98	720	0.69	250	2.00	166	3.01

N = number of parallel scratches; D = distance between scratches (μm).

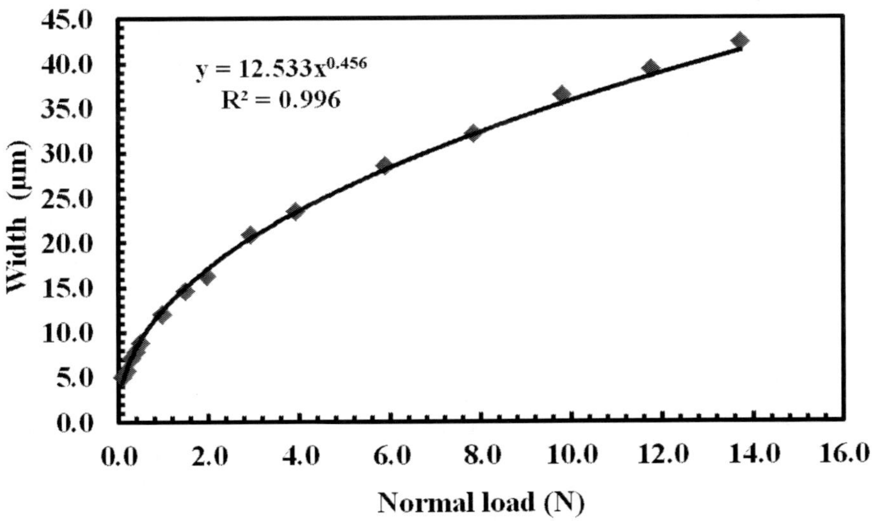

Figure 16. Indentation width as function of normal load for mono-indentations in M2 tool steel.

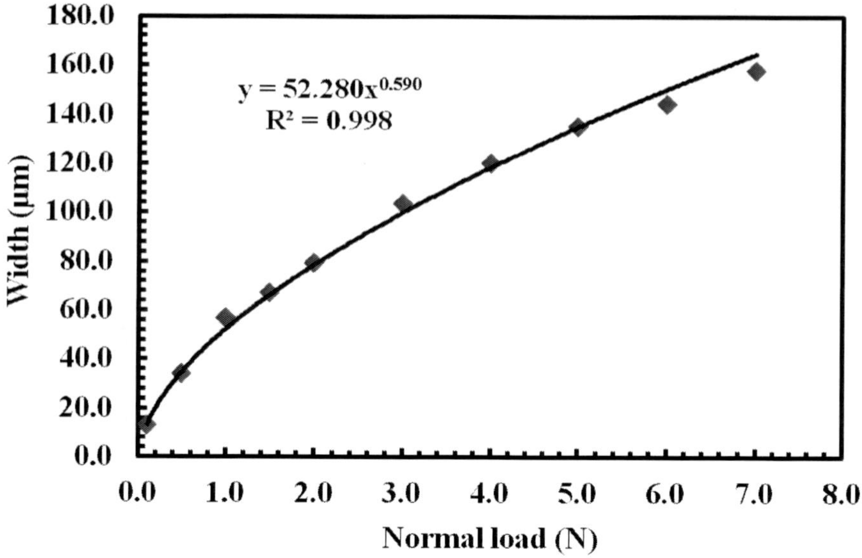

Figure 17. Effect of normal load on scratch width for individual scratches in copper samples.

A similar procedure was carried out to establish the relationship between normal load and scratch width, as shown in Figure 17 for copper. This relationship for M2 tool steel samples was shown previously in Figure 12.

Using the equations found in those experiments, the size of the events (indentation or scratch width) was well controlled by the normal load imposed. In the scratch superimposition tests, the degree of superimposition was controlled by knowing the normal load and the distance between scratches, according to Equation 3.

Figure 18 shows the lines of indentations obtained when using a normal load of 2.3 N on M2 tool steel sample.

In this test, the maximum length of the line of indentation was fixed as 300 μm and the number of indentations increased to vary the degree of superimpositions.

Figure 19 shows the topographical profiles of the lines marked in Figure 18, obtained using laser interferometry. When there is no superimposition between the indentations (Figure 19a), the depth of the events is constant, but it reduces when the number of indentations (NI) increases and then superimposition occurs (Figure 19b), since the material deformed by a new indentation partially covers the previous indentation.

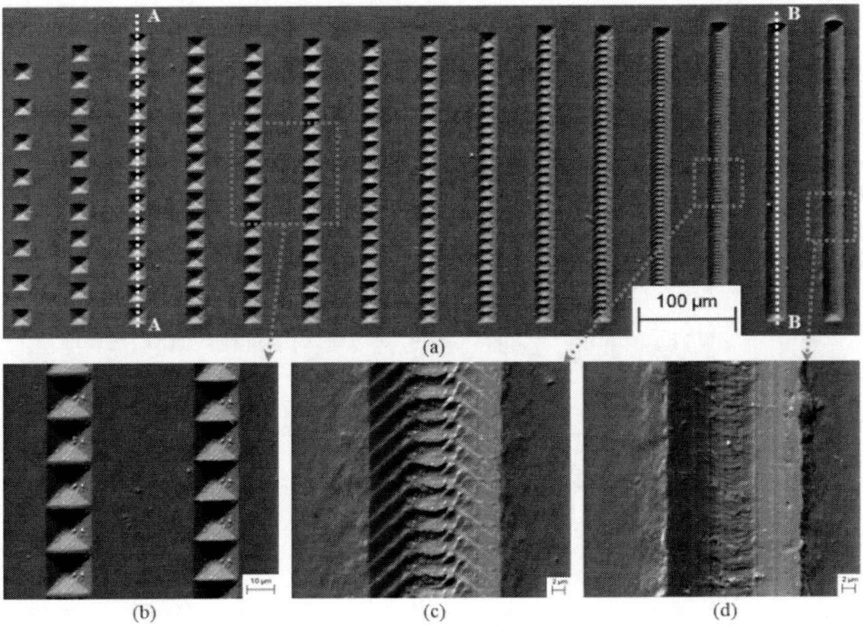

Figure 18. (a) Lines of indentations increasing the density of indentation. Superimposition degree: (b) 0% and 10%; (c) 80%; (d) 98%. M2 tool steel sample. SEM.

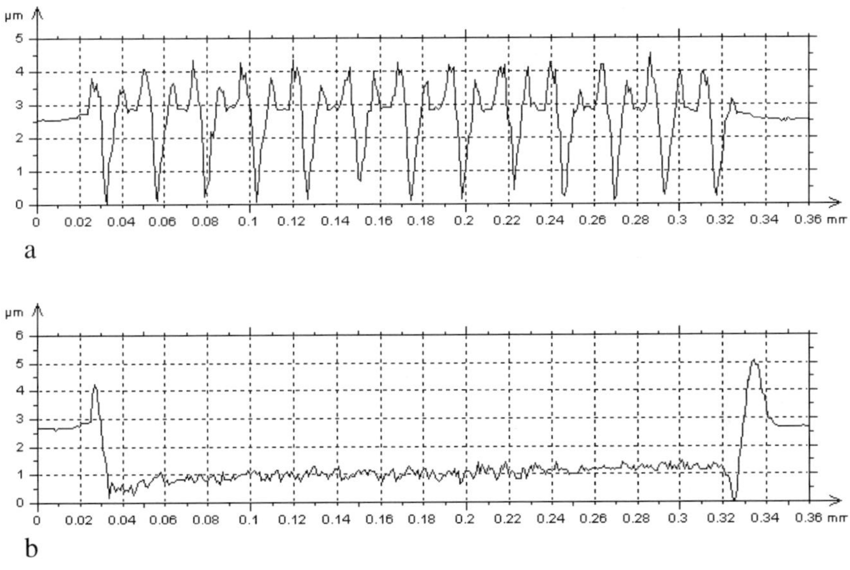

Figure 19. Laser interferometry of lines of indentations in tool steel; (a) profile AA; (b) profile BB.

Figure 20a shows the methodology used to measure the average depth (d_1, d_2...) below the initial reference line when superimposition occurs. The areas above and below the reference line were also computed (Figure 20b). These values are shown in Figure 21 as a function of both the degree of superimposition and the distance between indentations. The values above and below the reference line are very similar until a degree of superimposition of around 50%. In this region, probably the material deformed by the indentations was displaced to the sides of the events. For superimpositions larger than 50%, the area below the reference line increased sharply, suggesting material removal caused by intense repeated deformation. Figure 22 shows a schematic model for the superimposition of indentations.

The deformation caused by one indentation affects the region around and below it, as well reported in the literature [28]. The high stresses imposed during the indentation lead to work-hardening in the region represented by the grey area in Figure 22a. Also, for a ductile material, the indentation produced on the surface does not reproduce the exact form of the indenter due to the elastic/plastic response of the material.

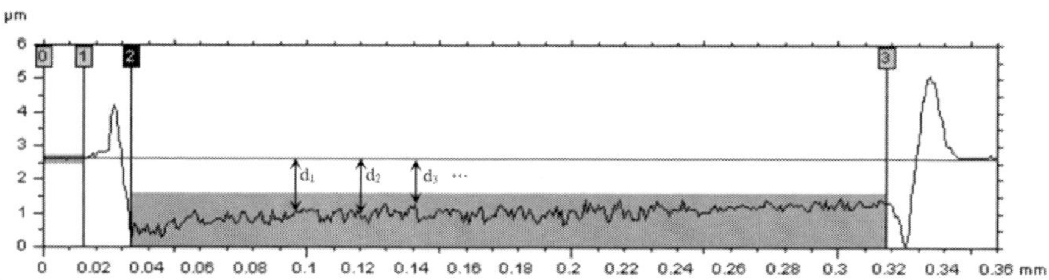

a

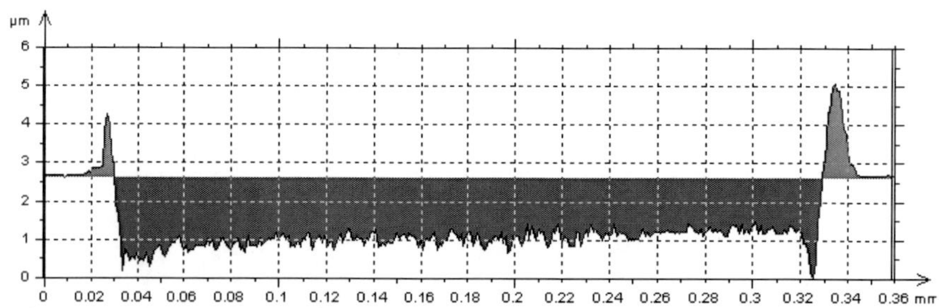

b

Figure 20. (a) Representation of the reference line and indentation depth when superimposition occurs and (b) areas of deformed material above and below the reference line for a line of indentations.

These phenomena are known in literature by “sink-in” and “pile-up” [29]. The present model does not take such phenomena into account.

The second indentation, at a distance DI, is performed over the work-hardened material left by the first indentation, as shown in Figure 22b. This material is also deformed during the second indentation. A portion of the deformed material during the second indentation is displaced into the mark left by the first indentation, as shown in Figure 22b. This hypothesis is corroborated by the results shown in Figure 18c and d.

Indeed, the empty region in the first indentation decreases the mechanical support for the second indentation. This lack of mechanical support suggests that during subsequent interactions the material can be strongly deformed, leading to shear-tongue formation and, as a consequence, to wear.

This model is valid for superimposed indentations. The result presented in Figure 21 indicates that for degrees of superimposition higher than 50% the material was strongly deformed to the sides of indentation lines and, possibly, removed in the region of low mechanical support.

Figure 18c and d show a strong deformed region in the centre of the indentation line for a high degree of superimposition.

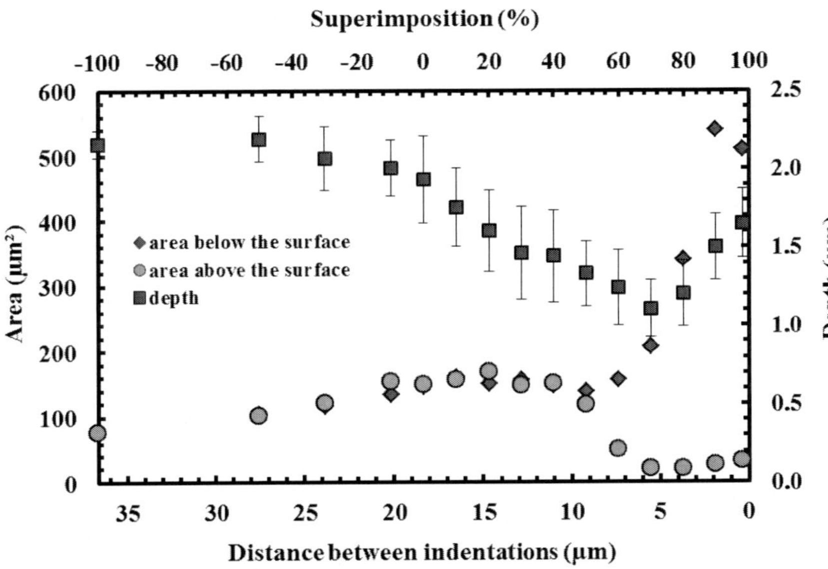

Figure 21. Variation of the areas above and below the reference line with the degrees of superimposi-tion for lines of indentations.

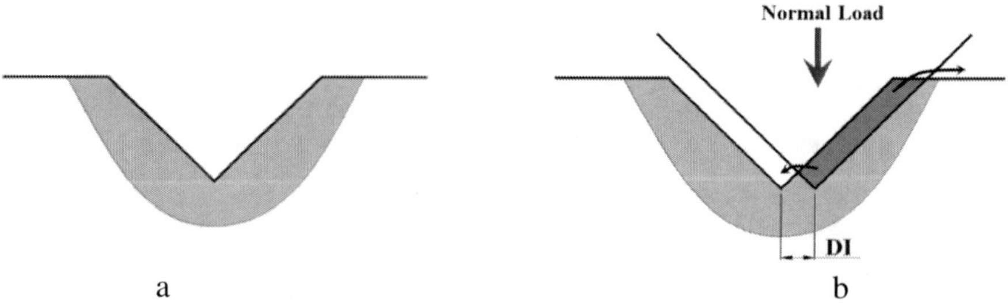

Figure 22. Indentation superimposition model. (a) schematic cross section of a single indentation; b) a second indentation at a distance DI from the first indentation.

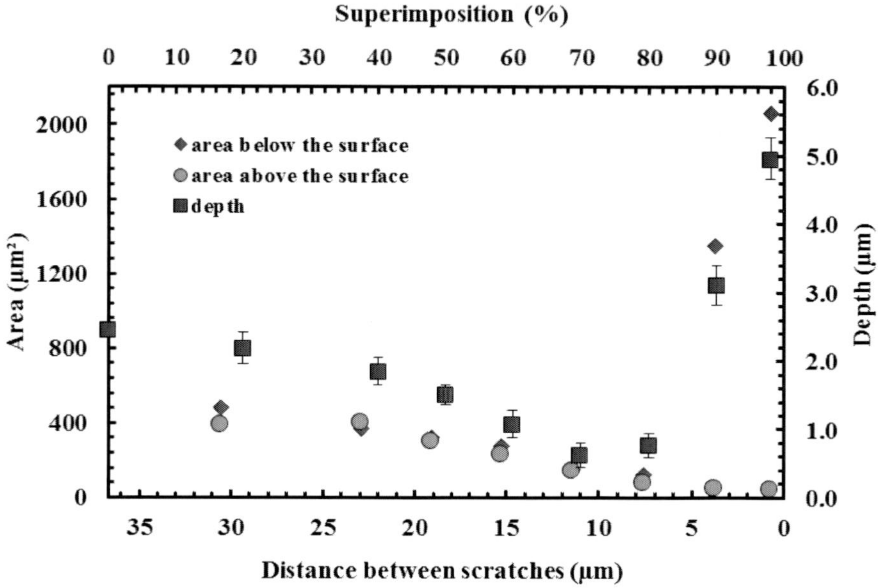

Figure 23. Variation of the areas above and below the reference surface with the degree of superimposi-tion for parallel scratches on M2 tool steel.

These results strongly indicate that it is possible to simulate material removal in ductile materials using superimposed indentation tests when the particle dynamics in the system to be simulated involves rolling and therefore repeated indentation of the surface.

For parallel scratches, we used in this work the concept of an average profile, which represents the mean value of the depth of each individual scratch on the surface.

The mean depth for the average profile and the areas above and below the reference line were computed and are shown in Figure 23 as a function of both the degree of superimposition and the distance between scratches.

Increasing the degree of superimposition, the mean depth of the average profile decreased slightly until 80% of superimposition. After that, an increase of the degree of superimposition resulted in a significant increase in the mean depth of the average profile. Also, the values of area above and below the surface were very similar up to 80% of superimposition, suggesting plastic deformation without significant material loss. Higher degrees of superimposition lead to material loss, since the volume of removed material, which is given by the void volume of material below the surface, showed a strong increase. Figure 24 shows examples of some selected SEM micrographs of scratches produced in tool steel samples for a normal load of 5.9 N (intermediate load) and different degrees of superimposition. When superimposition of the scratches did not occur (Figure 24a), microploughing combined with some chip formation at the sides of the scratches were present. As reported by Gore and Gates [9], this wear mechanism is intermediate between microploughing and microcutting, showing deformed material and small chips at the side of the scratch. With the increase of the degree of superimposition, microploughing was reduced and chip formation became very severe, Figure 24c and d. For very high degrees of superimposition, removal of material became evident by the great quantity of debris found in Figure 24d, reinforcing the idea that high levels of superimposition are required to cause wear. In order to support the comments above, some selected SEM images of the cross sections of the simulated samples are shown in Figure 25. The dotted line in (a) shows schematically the procedure used to calculate the f_{ab} factor, given by $f_{ab} = (A_B - A_A)/A_B$, where A_B = area below the reference line (dotted) and A_A = area above the reference line (dotted). A f_{ab} factor = 0 indicates pure microploughing, whereas a f_{ab} factor = 1 indicates pure microcutting [8]. In a previous work, Williams and Xie [18] showed that when reducing the distance between parallel scratches, the wear mechanism changed from ploughing to cutting. Such behaviour was attributed to the interaction between neighbouring scratches. Transitions in wear mechanisms were further investigated using scratching tests in copper. Figure 26 shows some selected micrographs of scratches produced in copper for different degrees of superimposition. In Figure 26a, there is no superimposition (-50%) and the interaction between the scratches is negligible. A considerable amount of plastic deformation around the scratches is observed, showing prevalence of microploughing.

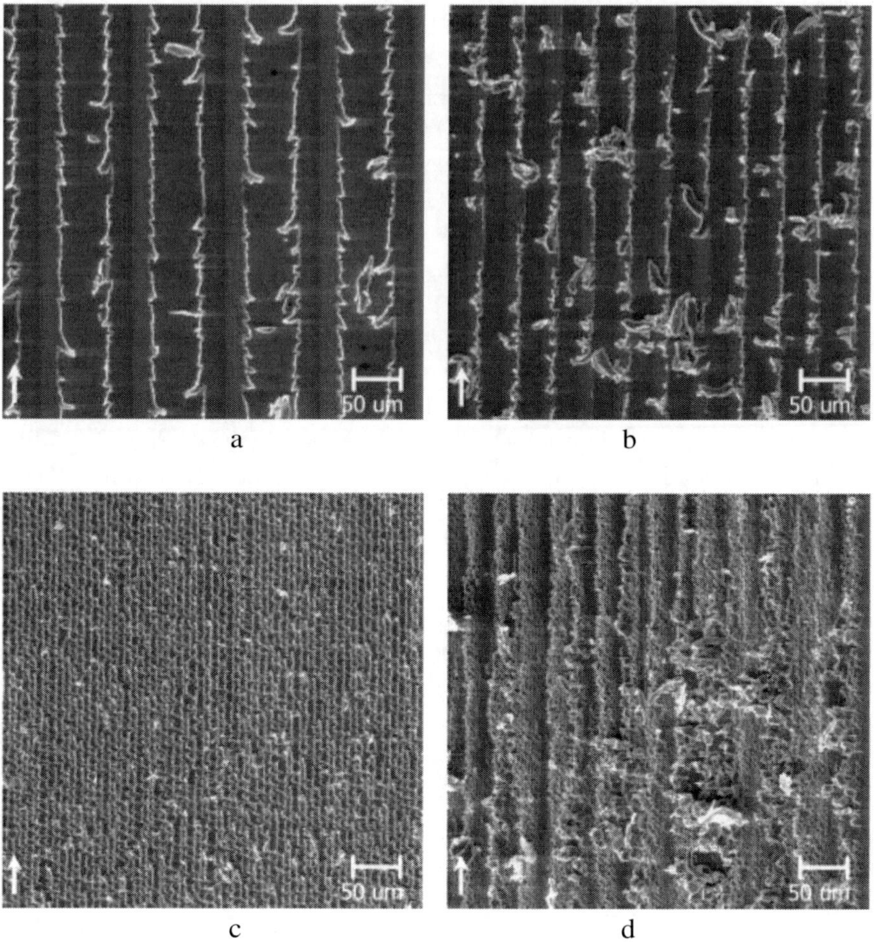

Figure 24. SEM of parallel scratches in tool steel for a normal load of 5.9 N; (a) DS = -100%; (b) DS = 20%; (c) DS = 85%; (d) DS = 95%.

The material is primarily pushed to the sides of the scratches, with very small loss of material, as also observed by Mezlini et al. [20] for the simulation of abrasive wear in annealed Al alloys. Increasing the degree of superimposition, a transition in wear mechanism occurs. For high degrees of superimposition, some kind of chip formation occurs, which is eventually removed, leading to material removal, as confirmed by Figure 27, where wear debris (indicated by arrows) located near the final end of the parallel scratches have a shape similar to the chips formed on the sides of the scratches.

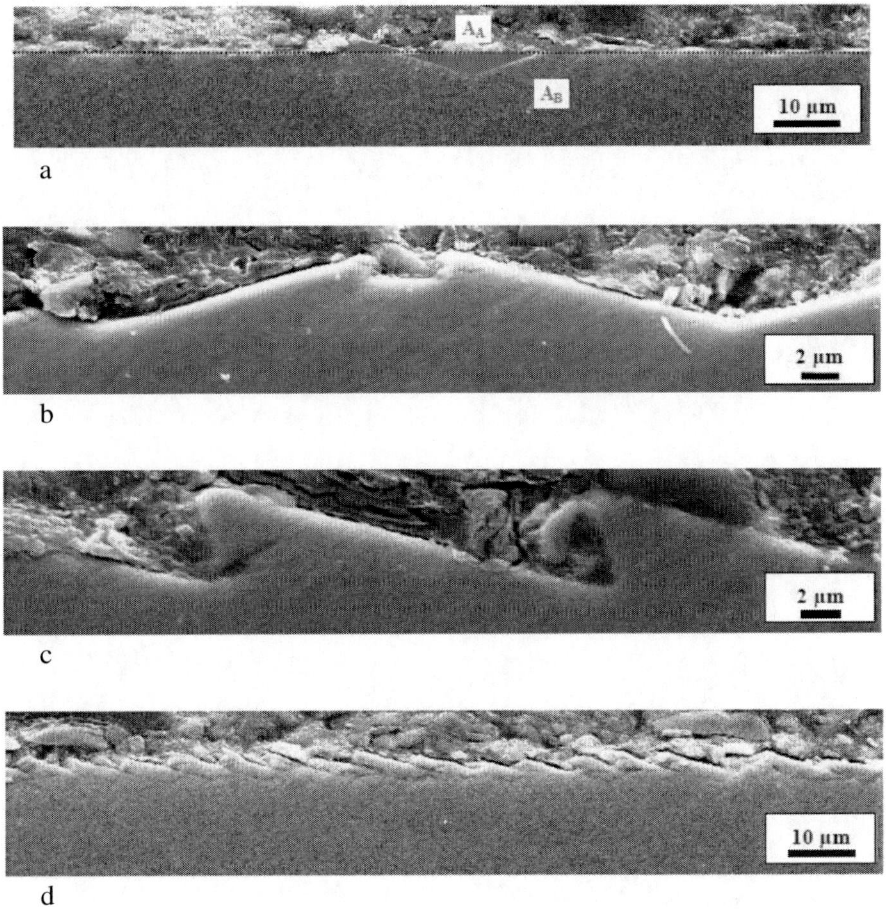

Figure 25. SEM cross sections of parallel scratches in tool steel for a normal load of 5.9 N; (a) isolated scratch; (b) DS = 0%; (c) DS = 50% and (d) DS = 80%.

Again, a transition in wear mechanism when the degree of superimposition was varied could be detected.

For an individual scratch the f_{ab} factor was calculated as 0.26, confirming the prevalence of microploughing.

With the increase of the degree of superimposition, some type of chip formation occurred. This was believed to be due to strain hardening of the sample caused by the interaction between neighbouring scratches, but possibly also to some geometrical effect at the sides of the scratches, which have less mechanical support when they interact with neighbouring scratches.

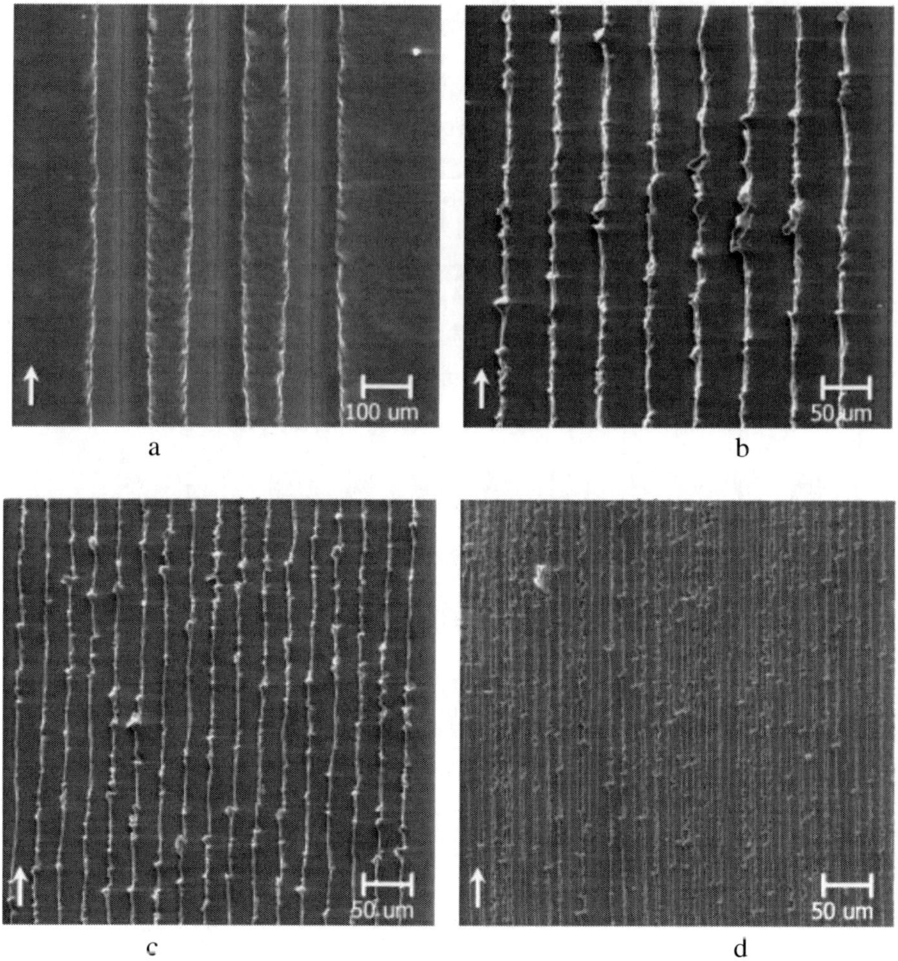

Figure 26. SEM of parallel groves in copper, normal load = 3 N; (a) DS = -50%; (b) DS = 60%; (c) DS = 80%; (d) DS = 95%.

In order to support the comments above, Figure 28 shows selected cross sections of the copper samples used in the simulations with parallel scratching. To investigate the increase of strain hardening when superimposition increases, these cross sections were etched. They show evidence of strong plastic deformation near the worn surface.

Each individual scratch becomes smaller and more deformed as the degree of superimposition increases.

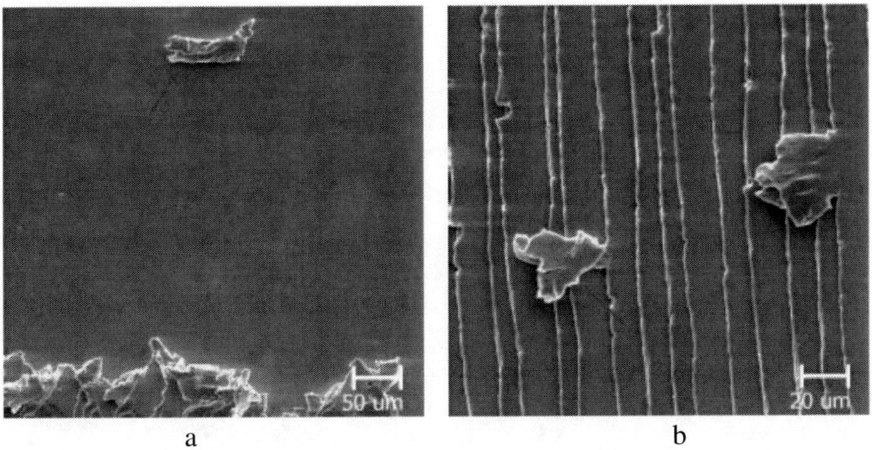

Figure 27. Wear debris near the scratching area (a) and inside the area (b), copper sample, normal load = 3 N, DS = 90%.

This is effect is so pronounced for DS = 95% (Figure 28d) that the individual scratches appear as small dents in the cross section and only the last scratch, which was not superimposed by a subsequent one, is properly visualized.

Figure 29a shows microhardness profiles for the copper samples tested with different degrees of superimposition. Figure 29b shows as an example the micrograph with microhardness measurements for a high degree of superimposition (98%).

Microhardness profiles show a marked increase in hardness near the deformed surface. This result is attributed to the work-hardening caused by the multiple interaction process.

The variation of the microhardness near the surface with the degree of superimposition is summarized in Figure 30. It evidences a tendency of increase of the subsurface microhardness as a function of the degree of superimposition, in particular for the higher degrees of superimposition. This result indicates that work-hardening induced by surface deformation is affected by the interaction between parallel scratches.

As discussed above, higher degrees of superimposition lead to higher amounts of removed material, which must be associated to accumulation of strain within the material.

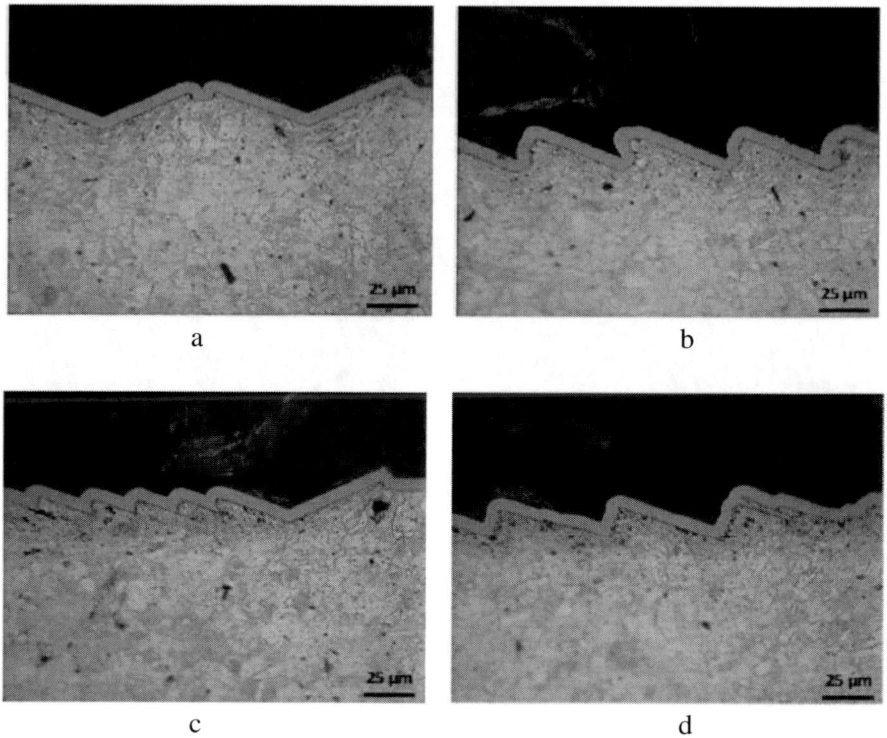

Figure 28. Etching of the cross sections of copper samples for a normal load of 3 N. (a) DS = 0%; (b) DS = 50%; (c) DS = 80%; (d) DS = 95%. OM.

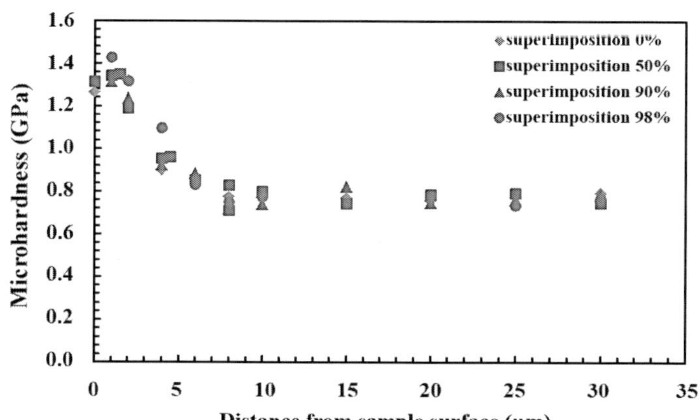

a

Figure 29. (Continued).

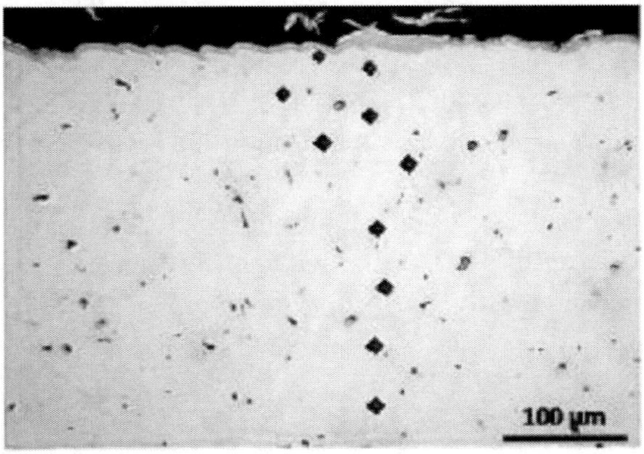

b

Figure 29. (a) Microhardness profiles for different degrees of superimposition; (b) example of micro-graph showing microhardness measurements, DS = 98%.

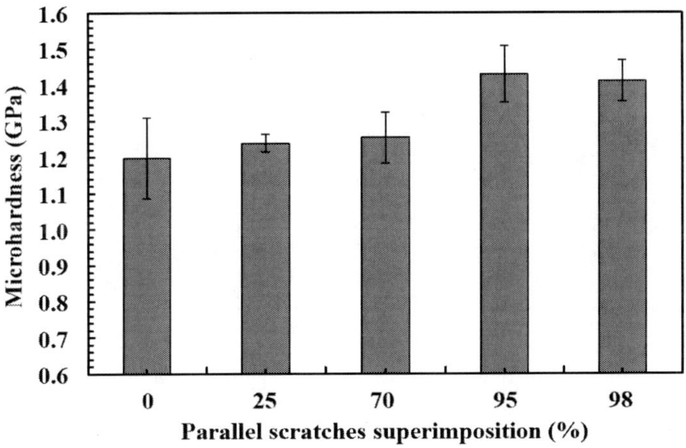

Figure 30. Variation of microhardness near the surface with the degree of superimposition.

2.3. Random Indentation Tests in Brittle Materials

In another approach, indentations randomly positioned on a soda-lime glass surface were carried out to investigate further the effect of the superimposition of indentations in a brittle material.

The simulated area was 0.5 x 0.5 mm using eight levels of normal forces: 0.15 N; 0.29 N; 0.44 N; 0.49 N; 0.59 N; 0.98 N; 1.47 N; and 1.96 N; and six levels of number of events: 500; 1000; 1500; 2000; 3500; and 5000. A Vickers indenter was used and the normal indentation load was kept constant during the simulations.

Figure 31 (laser interferometry) and Figure 32 (SEM) show selected areas obtained in these tests. The tests with lower normal loads and fewer indentations produced mostly plastic deformation.

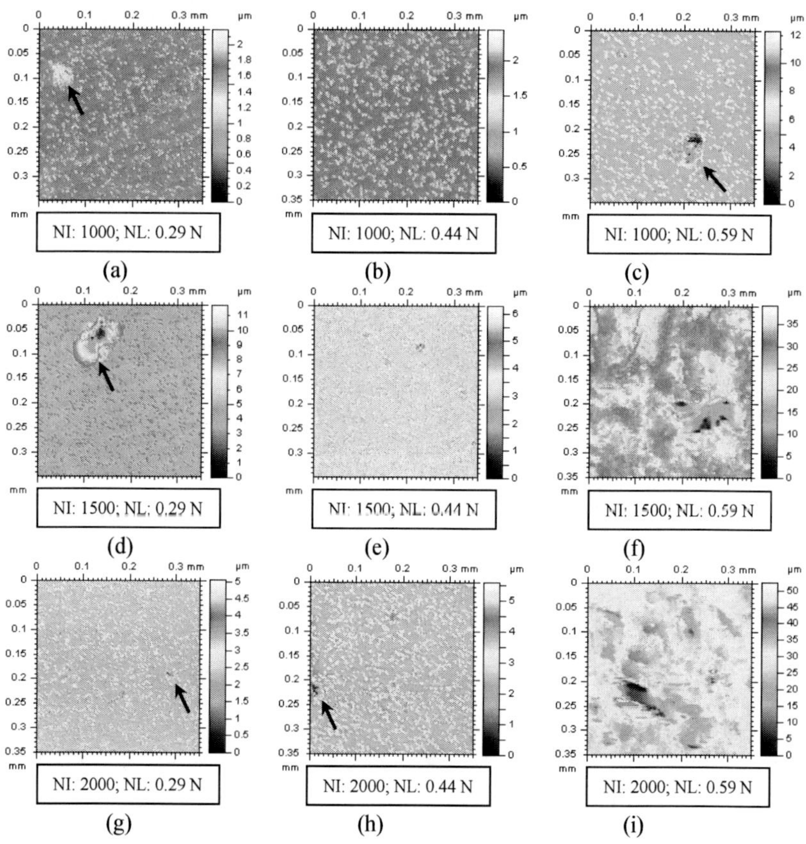

Figure 31. Areas with random indentations in soda-lime glass for different loads (NL) and number of indentations (NI), LI.

On the other hand, in areas where the degree of superimposition was large, crack propagation led to material removal.

Therefore, large degrees of superimpositions resulted in the occurrence of spalling, as indicated by the arrows. This probably occurs due to the proximity of stress fields of the indentations.

With the increase of the number of indentations and of the normal load, superimposition was more frequent and therefore spalling more widespread over the surface.

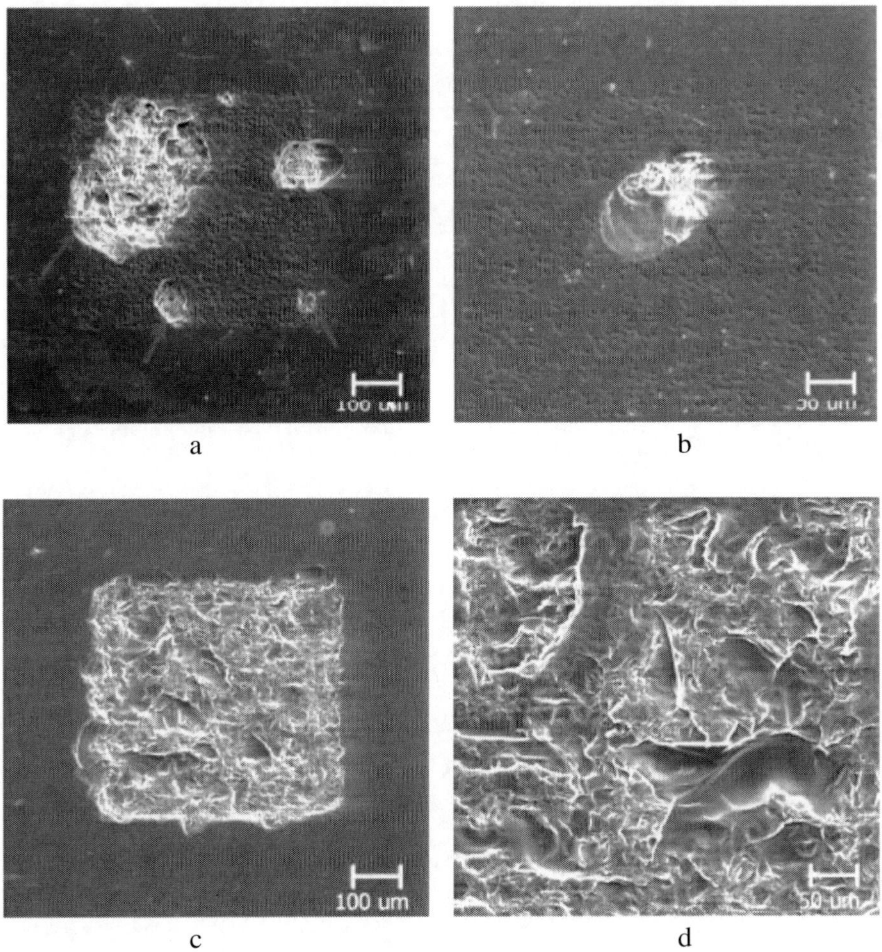

Figure 32. SEM of randomly positioned indentations in soda-lime glass samples; (a) and (b) 1500 indentations, 0.29 N; (c) and (d) 2000 indentations, 0.59 N.

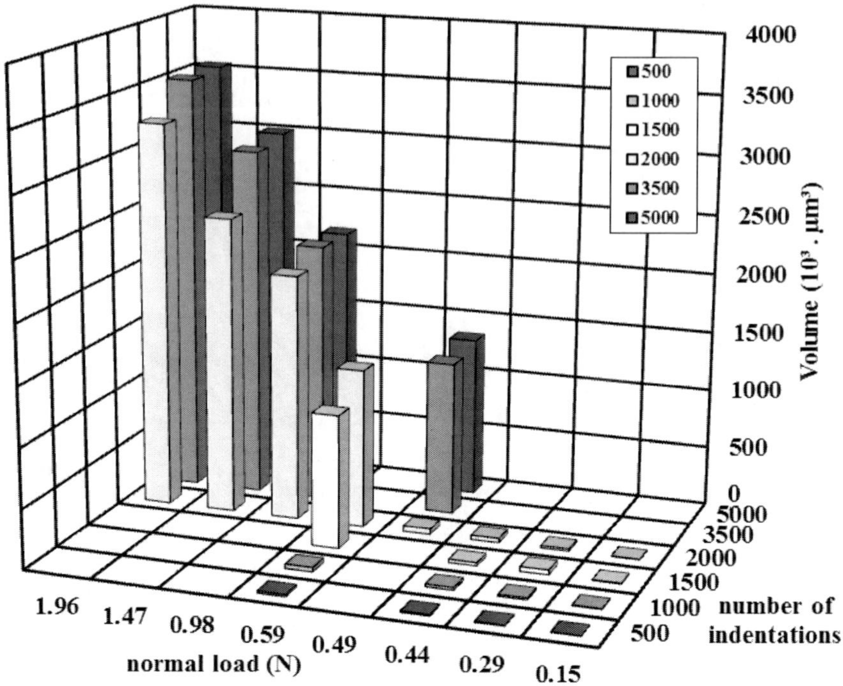

Figure 33. Mapping of wear volume as a function of normal load and number of indentations for soda-lime glass.

Figure 33 shows the wear volume for each test as a function of the normal load and of the number of indentations. The calculation of wear volumes was carried out using the surface topography measurements by LI. The area measured by LI was slightly larger than the whole simulated area, so that the area outside the simulated area could be used as the reference surface. The total volume of valleys in relation to the reference surface was then defined as the wear volume.

This figure shows a clear transition from mild wear, with small occurrence of microcracking, to severe wear, with intense microcracking, due to increase of the superimposition of indentations. The onset from mild wear (ductile) to severe wear (brittle) increased the wear volume in around 2700%.

CONCLUSION

In this chapter the mechanisms that result in material removal by abrasive particles were discussed considering the particle dynamics at the interface. The abrasive wear mechanisms were reproduced using very well controlled scratch tests or indentation tests. The main idea is that the sliding of abrasive particles was simulated by series of scratches and the rolling of abrasive particles was simulated by series of indentations.

This work has clearly identified the effect of the superimposition of individual particle interactions with the surface (mono-events) during abrasive wear. Whereas individual events can result in certain wear mechanisms, the superimposition of those events can lead to drastic changes in wear mechanisms, and, as a result, surface morphology and topography can be drastically changed. A new parameter for the simulation of abrasive wear was then proposed, called degree of superimposition.

Results of experimental simulations were presented for annealed copper, tool steel and glass. Superimposed scratches on copper showed clearly that mono-interactions resulted mainly in plastic deformation (microploughing), but that the increase of the degree of superimposition resulted in mass loss (microcutting). Superimposed indentations and superimposed scractches in tool steel also resulted in mass loss for large degrees of superimposition. For all the ductile materials tested, work hardening was intensified for high degrees of superimposition, and the superimposition of the events reduced the support for subsequent events, increasing mass removal for higher degrees of superimposition. For glass samples, the main transition observed was between ductile and brittle abrasion. Randomly distributed indentations showed that the increase of normal load and degree of superimposition promoted severe spalling. It is also relevant to point out that the experimental simulation stores information about all the forces associated to each individual event. The use of the information obtained with the simulation here proposed, where the precise location of each event, the forces involved and the depth resulting of the interaction of the abrasive particle with the sample surface are precisely known, should lead to the development of a new and more realistic computational simulation approach of abrasive wear.

In conclusion this chapter proposed a new methodology to experimentally simulate abrasive wear that involves well-controlled instrumented indentation and/or scratch tests, where indentations represent the contact of rolling abrasive particles without sliding, and scratches represent the sliding of abrasive particles.

The experimental simulation here proposed was able to reproduce real worn surfaces, both topographically and morphologically. The results showed that the superimposition of individual events in the simulation had a great effect on the morphology and topography of the simulated surfaces and could result in wear mechanism transitions.

REFERENCES

[1] Adachi, K., Hutchings, I. M. (2003). Wear mode mapping for the micro-scale abrasion test. *Wear*, 255, 23-29.

[2] Burwell, J. T. (1957). Survey of Possible Wear Mechanisms. *Wear*, 1, 119-141.

[3] Rutherford, K. L., Hutchings, I. M. (1996). A Micro-abrasive Wear Test, With Particular Application to Coated Systems. *Surface and Coatings Technology*, 79, 231-239.

[4] Rutherford, K. L., Hutchings, I. M. (1997). Theory and Application of a Micro-Scale Abrasive Wear Test. *Journal of Testing and Evaluation*, 25, 250-260.

[5] Trezona, R. I., Allsopp, D. N., Hutchings, I. M. (1999). Transition between two-body and three-body abrasive wear: influence of test conditions in the microscale abrasive wear test. *Wear*, 225-229, 205-214.

[6] Goddard, J., Wilman, H. (1962). A theory of friction and wear during the abrasion of metals. *Wear*, 5 (2), 114-135.

[7] Sedriks, A. J., Mulhearn, T. O. (1963). Mechanics of Cutting and Rubbing in Simulated Abrasive Process. *Wear*, 6, 457-466.

[8] Zum Gahr, K. L. (1987). Microstructure and Wear of Materials. *Tribology Series,* Volume 10, Elsevier Science Publishers B. V., Amsterdam, p.560.

[9] Gore, G. J., Gates, J. D. (1997). Effect of hardness on three very different forms of wear. *Wear*, 203-204, 544-563.

[10] Kayaba, T., Hokkirigawa, K., Kato, K. (1986). Analysis of the abrasive wear mechanism by successive observations of wear processes in a scanning electron microscope. *Wear*, 110 (3-4), 419-430.

[11] Williams, J. A., Hyncica, A. M. (1992). Mechanisms of abrasive wear in lubricated contacts. *Wear*, 152, 57-74.

[12] Williams, J. A., Hyncica, A. M. (1992). Abrasive Wear in Lubricated Contacts. *Journal of Physics D-Applied Physics*, 25 (1A), A81-A90.

[13] Buijs, M., Korpel-van Houten, K. (1993). Three-body abrasion of brittle materials as studied by lapping. *Wear*, 166, 237-245.

[14] Verspui, M. A., With, G., van Der Varst, P. G. Th., Buijs, M. (1995). Bed thickness and particle size distribution in three-body abrasion. *Wear*, 188, 102-107.

[15] Fang, L., Kong, X. L., Su, J. Y., Zhou, Q. D. (1992). A Wear Tester Capable of Monitoring and Evaluating the Movement Pattern of Abrasive Particles in 3-Body Abrasion. *Wear*, 159(1), 115-120.

[16] Fang, L., Kong, X. L., Su, J. Y., Zhou, Q. D. (1993). Movement Patterns of Abrasive Particle in Three-body Abrasion. *Wear*, 162-164, 782-789.

[17] Bozzi, A. C., de Mello, J. D. B. (1999). Wear Resistance and Wear Mechanisms of Wc–12%Co Thermal Sprayed Coatings in Three-Body Abrasion. *Wear*, 233-235, 575-587.

[18] Williams, J. A., Xie, Y. (1992). The Generation of Wear Surfaces by the Interaction of Parallel Grooves. *Wear*, 155(2), 363-379.

[19] Williams, J. A. (1999). Wear Modelling: Analytical Computational and Mapping: A Continuum Mechanics Approach. *Wear*, 225-229, 1-17.

[20] Mezlini, S., Kapsa, Ph., Henon, C., Guillemenet, J. (2004). Abrasion of aluminium alloy: effect of subsurface hardness and scratch interaction simulation. *Wear*, 257 (9-10), 892-900.

[21] Da Silva, W. M., de Mello, J. D. B. (2006). Simulação de Desgaste Abrasivo via Eventos Múltiplos. *Ciência and Tecnologia dos Materiais* (ISSN: 0870-8312), 18 (1/2), 49-56 (in Portuguese).

[22] Da Silva, W. M., de Mello, J. D. B. (2009). Using parallel scratches to simulate abrasive wear. *Wear*, 267, 1987-1997.

[23] Da Silva, W. M., Costa, H. L., de Mello, J. D. B. (2009). Superimposition of Interactions: a New Approach to Simulating Abrasive Wear Mechanisms. *WTC IV - World Tribology Congress 2009*, September, Kyoto, Japan.

[24] Labiapari, W. S., Alcântara, C. M., Costa, H. L., de Mello J. D. B. (2013). Stainless steel as an antiwear material for the bio-fuel industry, article in press. *Wear*, http://dx.doi.org/10.1016/j.wear.2012.12.023.

[25] Stout, K. J., Blunt, L., Dong, W. P., Mainsah, E., Luo, N., Mathia, T., Sullivan, P. J., Zahouani, H. (2000). Development of Methods for the Characterisation of Roughness in Three Dimensions. *Butterworth-Heinemann,* ISBN 1857180232, chapter 12.

[26] Whitehouse, D. J. (1974). The measurement and analysis of surfaces. *Tribology International*, Volume 7, Issue 6, (December 1974), 249-259.

[27] ISO 4287 (1997). Geometrical Product Specifications (GPS) – Surface Texture: Profile Method – Terms, Definitions and Surface Texture Parameters. *International Standard ISO 4287,* first edition, published in April 1st of 1997.

[28] Jonson, K. L. (1985). *Contact Mechanics*. Cambridge University Press, Cambridge, p. 241.

[29] Oliver, W. C., Pharr, G. M. (2004). Measurement of hardness and elastic modulus by instrumented indentation: Advances in understanding and refinements to methodology. *Journal of Materials Research*, 19, No 1, 3-20.

In: Recent Advances in Abrasives Research ISBN: 978-1-62808-566-2
Editor: Dirk Bähre

Chapter 4

THE CONCEPT OF ENERGETIC INTERACTION BETWEEN THE TOOL AND THE WORKPIECE IN SUPERFINISHING PROCESSES

A. N. Tyurin[1]
West-Kazakhstan Agrarian-Technical University
named after Zhangir Khan, City of Uralsk, Republic of Kazakhstan

ABSTRACT

The energy parameters of the superfinishing process, and, in particular, the one-time cutting energy, are mathematically related to such important parameters as the quantity of metal removed during processing and the roughness of the finished surface. The one-time cutting energy, on the one hand, is determined by the area of the individual cutting operations, and, consequently, the depth of the residual scratches on the surface of the workpiece and the obtained surface roughness; on the other hand, this criterion depends on the length of the individual cuts, and thus, together with the depth of the cuts, characterizes the intensity of stock removal. An increase in cutting speed leads to an increase in one-time cutting energy at a constant depth of cut, and consequently, this provides conditions to increase processing capacity without the need to increase

[1]Corresponding author. Email: turin56@mail.ru.

the roughness of the surface to be machined.

The energy-related approach to modelling the superfinishing process provides the option to create a more universal model of the process, which allows to search for new ways to improve the technology and the control of the technological process of final precision surface processing of a wide range of parts.

Keywords: Superfinishing, material removal, roughness, cutting energy

INTRODUCTION

Currently, there is a number of different theoretical approaches to modelling the process of superfinishing. However, they all have a predominantly individual character, reflecting specific conditions of the superfinishing process, which limits their practical use in improving the superfinishing technology in general.

The essence of all processes of the mechanical processing of workpieces, including the process of superfinishing, lies in the process of converting energy from one form to another. During superfinishing, the energy of the equipment drive is converted to the cutting energy of individual abrasive grains and to the friction energy of the chips and sludge on the surface being machined.

Thus, the superfinishing process is, above all, an energetic process, and so, without a serious analysis of the energy parameters of the process, it is not possible to draw a sufficiently full picture of the nature of this process. Cutting energy and friction energy is converted partly into internal energy of the workpiece and binding energy of the tool, and partly into thermal energy. The thermal energy released during the process is accompanied by microchipping, and the growth of the internal energy is accompanied by a hardening of the surface layer of the workpiece. The increase of the internal binding energy holding the grains on the surface of the instrument, is accompanied by its destruction, resulting in, on the one hand, a self-sharpening of the tool and an increase of its cutting capacity, and, on the other hand, a decrease in the depth of individual cuts, accompanied by a reduction in the rate of stock removal and a reduction of the roughness of the finished surface.

All energy parameters of the superfinishing process can be divided into individual and generalized parameters. The individual parameters reflect the impact of individual abrasive grains on the surface to be machined.

The most common individual parameter is the one-time cutting energy of a single cut. By the one-time energy of a single cut we understand the energy expended during a continuous cut of workpiece material by a single abrasive grain.

This cut may be complete or incomplete. A complete cut is obtained when a single abrasive grain gets out of contact with the surface to be machined without damage. Such cases of cuts are usually unlikely as they are obtained with a very small depth of cut of the abrasive grains that are strongly held by the binding energy. However, it is known that during superfinishing the stock is actively removed under the condition of an active self-sharpening of the abrasive tool. For this purpose, it is necessary to provide machining conditions, under which the individual abrasive grains actively drop out of the bond of the tool under the effect of the surface to be machined. If an abrasive grain in continuous contact with the surface to be machined drops out of the bond and stops cutting, we call this an incomplete cut. This decreases both the depth and length of the cut which leads to a reduction of the one-time energy of a single cut.

Therefore, the one-time energy of a single cut depends on the one-time depth of cut and one-time length of cut. And although the parameters of the cutting part of the abrasive grains depend on the material, it applies that for each material, the medium-probable parameters of the cutting part of the grains depend on the depth of cut so that the relation between the depth and width of cuts remains constant. Therefore, instead of the two individual energy parameters of the superfinishing process – the one-time depth of cut and the one-time area of cut – we have adopted only one parameter: the one-time depth of a single cut.

The one-time depth of cut shall mean the medium-probable depth of a cut with a single abrasive grain. The one-time length of cut shall mean the medium-probable length of a continuous cut with a single abrasive grain. The word "one-time" highlights that, in the case at hand, we examine a single cut performed by a single abrasive grain during one continuous contact with the surface to be machined.

Since the energy of a cut depends not only on the parameters of the cut itself, also the internal energy of the machined material of the workpiece is to be considered among the individual energy parameters. The internal energy of the machined material of the workpiece shall mean the energy required to remove volume units of material from the surface to be machined with a single abrasive grain. The product of the volume of a single cut and the internal energy of the machined material should be equal to the one-time energy of a single cut.

Since, as noted above, single abrasive grains perform mostly incomplete cuts during superfinishing, one of the important individual energy parameters is the internal energy of the bonding material. The internal energy of the bonding material shall be the average likely amount of deformation energy of the bond posts at which the grains drop out of the tool's bond. Since the grain retention strength of the tool's bonding material and the probability of grain peaks above the level of the bonding material depends on the distance between the grain peaks and the level of the bonding material, the internal energy of the bonding material will depend on the distance between the level of the bonding material and the surface to be machined during the finishing process. Thus, the given energy parameter depends on the processing conditions, mainly on the characteristics of the tool and the specific force used to press the tool against the workpiece.

Among the general energy parameters of the superfinishing process are the specific cutting force and the specific frictional force. To the specific cutting force we assign that part of the force used for the performance of the superfinishing process which is consumed per unit of time for the performance of single cuts performed by abrasive grains that are in simultaneous contact per unit of area of the surface to be machined. This item is also medium-probable as it depends on the medium-probable value of the one-time energy of individual cuts and on the medium-probable number of active abrasive grains per unit of the surface to be machined. If the specific cutting force is multiplied by the contact area of the tool and the workpiece, we get the cutting force.

The specific power of friction is the medium-probable part of the force used to implement the superfinishing process which is consumed by the friction of the chips and sludge within the pores of the abrasive tool from one unit of the workpiece's surface to be machined. In our opinion, this is one of the most important energy parameters of the superfinishing process. On the one hand, the higher the specific frictional force, the lower the efficiency of the process of the technological system and the lower the rate of stock removal. But on the other hand, the higher the specific frictional force, the less the depth of penetration of the abrasive grains into the surface of the workpiece and the lower the roughness of the finished surface. It is known that without an active clogging of the working surface of the abrasive tool in the final part of the operation, it is not possible to get low roughness values of the surface to be machined, even if a very fine-grained tool is used. Therefore, it is very important to keep an effective balance between the cutting force and the specific frictional force in the initial and final stages of the superfinishing process.

The quality of the surface layer of the finished surface depends on the additional internal energy, which is applied to the surface to be machined during the process of plastic deformation. The increase in this energy is determined by the normal cutting force of a single grain and the depth of deformation, which corresponds to the one-time depth of cut.

This article discusses the parameter of a single cut of an abrasive grain during abrasive machining with a stone – in this, we will consider the general cutting parameter as well as the cutting force, and the depth, length and volume of a single cut. Each of these parameters characterizes the cut from one point of view, and the parameter of the energy of a single cut characterizes the nature of a cut with an abrasive grain in general.

ONE-TIME ENERGY OF CUTTING BY SINGLE GRAIN

The work (in joules, J) expended for cutting with one grain within an elementary time interval, can be found from the expression:

$$dA = P_z \cdot \vec{v}_r \cdot dt \tag{1}$$

where is

$\vec{v}_r$ – cutting speed (m/s)

P_z – tangential cutting force in the direction of the cutting speed, *(H)*.

Cutting forces during abrasive treatment were investigated by many authors [1, 2], and others. For example, S. Korczak [3] found that the cutting forces with a single abrasive grain are:

$$P_y = \frac{\sqrt{3{,}25} \cdot a \cdot \tau_s}{\sin\beta_1} \cdot \sin\beta + 0{,}5 \cdot \tau_s \cdot l_z;$$

$$P_z = \frac{\sqrt{3{,}25} \cdot a \cdot \tau_s}{\sin\beta_1} \cdot \cos\beta + 0.5 \cdot \mu \cdot \tau_s \cdot l_z,$$

where is

a – the depth of cut, mm;

τ_s – tangential shear stress, MPa;

l_z – length of the wearing areas of the grain's working surface, mm;

μ – coefficient of sliding friction;

β_1 – angle of the shear plane of the material during the cutting process;

β – angle between the cutting speed vector and the shear plane.

These dependencies are most often used in the study of the grinding process. But, in our opinion, they are not suitable for the study of finishing processes with abrasive stones. The reason is that the above dependencies reflect a decisive influence of the wear of the grain tips under the effect of high temperatures and the low degree of dimensional wear of the tool on the cutting force during grinding. Therefore, one of the components of the cutting force is the friction of the wearing surface of the grain l_s on the surface to be machined. During finishing with an abrasive stone, the wear of the cutting part of the abrasive grains has only an insignificant effect since the cutting temperatures are low and the abrasive tool operates in a mode of intense self-sharpening. But if there is no wearing surface, then, as follows from the above dependency, there is no frictional force – which is contrary to both the modern concept of the nature of friction and common sense.

Perhaps the most accurate model of the cutting forces during abrasive finishing is based on the dependencies proposed in [1]:

$$P_z = 34 \cdot \tau_s \cdot a^2 \cdot (1 + 1{,}5 \cdot \mu \cdot (1 + \xi));$$
$$P_y = 51 \cdot \tau_s \cdot a^2 \cdot (1 + \xi), \qquad (2)$$

where is

ξ – chip compression (relation between the length of cut and the chip length).

Apparently, the dependencies in (2) significantly differ from the equations shown above in the fact that the dependencies lack the length of the wearing area of the grains and that the squared depth of cut affects the cutting forces. The latter fact has been confirmed by numerous studies, for example, by S.G. Redko [2], who performed the static indentation of the working surface of the abrasive tool into the workpiece surface made of different materials. And although the static indentation of the grains into the surface to be machined by its physical nature differs from the cutting process, the nature of the dependency of the normal force on the depth of indentation is the same in both cases.

The speed of travel of the abrasive grain at a fixed point in time depends on the grain's path of travel. The grain's path of travel is determined by the kinematics of the finishing process and, in particular, the movement pattern of the abrasive stone. There are numerous methods of finishing using an abrasive stone, each featuring individual kinematics of the motion of the tool and the workpiece in relation to each other. But we will try to generalize the possible movement patterns of a single grain as much as possible.

The maximum depth of cut equals:

$$a_{zm} = h_z - h_b.$$

By substituting the maximum value of the depth of cut in equation (3)

$$a_z = h_z - h_b - \frac{1}{2} \cdot (r_b + h_b) \cdot \left(\frac{z_b}{H} \cdot \gamma_{bm} - \gamma_z\right)^2 (1-c) \text{ with } -z_{k1} \le z_b \le z_{bp} \qquad (3)$$

where is

h_z – value of the protrusion of the abrasive grain above the bonding material surface, mm;

h_b – distance between the location of the stone's bond level and the surface to be machined, mm.

H – travel range of the stone, *mm*,

z_b – coordinate of the centre of the stone, mm,

r_b – radius of the stone profile, mm.

We find the coordinate of the position of the stone at the maximum penetration depth of grain in the surface to be machined:

$$z_{bm} = \frac{H}{\gamma_{bm}} \cdot \gamma_z . \tag{4}$$

After the depth of cut (3) has been determined, it is easy to find the cutting force with formula (5)

$$\begin{aligned} P_z &= 34 \cdot \tau_s \cdot a^2 \cdot (1 + 1{,}5 \cdot \mu \cdot (1 + \xi)); \\ P_y &= 51 \cdot \tau_s \cdot a^2 \cdot (1 + \xi), \end{aligned} \tag{5}$$

However, to determine the energy expended by a single abrasive grain for cutting, we need to know the speed of travel of the grain relative to the surface to be machined.

It is known that during the process of abrasive finishing, the abrasive stone is communicated to perform an oscillating motion along the surface to be machined. These movements should be smooth to avoid the occurrence of strong dynamic forces during the finishing process, which would lead to vibrations of the technological system and failures of its mechanisms. Therefore, oscillating motions are usually communicated to the tool head by eccentric mechanisms, through the use of which the travel speed of the tool head decreases at the end of each turn, and then reverses, and during the movement of the tool head to the centre of the oscillating motion, its velocity increases and reaches its maximum value at the very centre.

In this case, the differential equation of the motion of the abrasive stone along the surface to be machined is as follows:

$$\frac{d^2 z_b}{dt^2} + 4 \cdot \pi^2 \cdot n_b^2 \cdot z_b , \tag{6}$$

where is

n_b – frequency of vibration of the stone, *Hz*.

Equation (6) is a second order linear differential equation with constant coefficients and without a constant term. According to the theory of linear differential equations, the general integral of this equation can be expressed as follows:

$$z_b = k_b \cdot sin(2 \cdot \pi \cdot n_b \cdot t + \psi),$$

where the items k_b and ψ have arbitrary values and are determined by the boundary conditions of the process. If the origin of ordinates is used as the zero-time reference t and the maximum stroke length of the stone is A, then

$$z_b = H \cdot sin(2 \cdot \pi \cdot n_b \cdot t). \tag{7}$$

Taking the derivative of expression (7) with respect to time, we find the velocity of the stone;

$$v_b = v_{bo} \cdot \cos(2 \cdot \pi \cdot n_b \cdot t); \tag{8}$$

where v_{bo} is the speed of travel of the stone in the centre of the profile of the workpiece to be machined, $м / c$, equaling

$$v_{bo} = 2 \cdot H \cdot \pi \cdot n_b.$$

The speed of travel of the stone and – together with it – the grain is only one component of the overall cutting speed. The second and main component of the cutting speed is the peripheral speed of rotation of the workpiece.

By adding these two speeds geometrically, we determine the total speed of travel of a single grain relative to the surface to be machined:

$$v_z = \pm\sqrt{(v_{bo} \cdot \cos(2 \cdot \pi \cdot n_b \cdot t))^2 + v_d^2} = \\ = \pm\sqrt{v_d^2 + v_{bo}^2} \cdot \sqrt{1 - \frac{v_{bo}^2 \cdot \sin^2(2 \cdot \pi \cdot n_b \cdot t)}{v_d^2 + v_{bo}^2}}; \tag{9}$$

where v_d is the peripheral speed of rotation of the workpiece, *m/s*.

Since the grain's depth of penetration into the surface of the workpiece a_z (3) depends on the position of the stone z_b relative to the surface to be machined, and the coordinate of the position of the centre of the stone z_b in turn depends on the time t, it follows that the depth of interaction of a single grain with the surface to be machined also depends on the time:

$$a_z = h_z - h_b - \frac{1}{2}\cdot(r_b + h_b)\cdot(\sin(2\cdot\pi\cdot n_b\cdot t)\cdot\gamma_{bm} - \gamma_z)^2(1-c) \tag{10}$$

$$\text{with} - t_{bn} \le t \le t_{bk}.$$

where t_{bn} and t_{bk} are, respectively, the start time of the kinematic contact with the abrasive grain with the surface to be machined and the end time of the contact.

By solving equations (9), (11) and (12)

$$\begin{aligned} P_z &= 34\cdot\tau_s\cdot a^2\cdot(1+1{,}5\cdot\mu\cdot(1+\xi)); \\ P_y &= 51\cdot\tau_s\cdot a^2\cdot(1+\xi), \end{aligned} \tag{11}$$

$$dA = P_z\cdot\vec{v}_r\cdot dt \tag{12}$$

expressing the abscissa of the position of the stone through time, we find:

$$\begin{aligned} dA = {} & 34\cdot\tau_s\cdot(1+1{,}5\cdot f\cdot(1+\xi))\cdot\sqrt{v_d^2+v_{bo}^2}\cdot(h_z-h_b)^2\times \\ & \times\left(1-\frac{r_b+h_b}{2\cdot(h_z-h_b)}\cdot(\sin(2\cdot\pi\cdot n_b\cdot t)\cdot\gamma_{bm}-\gamma_z)^2(1-c)\right)^2\times \\ & \times\sqrt{1-\frac{v_{bo}^2}{v_d^2+v_{bo}^2}\cdot\sin^2(2\cdot\pi\cdot n_b\cdot t)}\cdot dt. \end{aligned} \tag{13}$$

Integrating equation (13) by time, we determine the amount of energy expended by a single abrasive grain for cutting. In doing so, we can take into account that the grain penetrating the surface to be machined to some depth, exhausts its entire stock of internal bond energy (keeping it in the tool bond) and drops out of the tool surface. Or that the grain completely scratches its kinematically determined groove, and then drops out. Or that the grain scratches several grooves of different depths that increase each time. All of these options can be dealt with using a variety of integration limits for equation (13).

We assume that the grain is at a distance h_z from the bond level of the tool and at an angle γ_z relative to the plane of symmetry of the stone. Then, the time of its possible start of contact with the workpiece can be determined from the equations:

$$t_{kl} = \frac{1}{2\cdot\pi\cdot n_b}\cdot\arcsin\left(\left(-\sqrt{\frac{2\cdot(h_z-h_b)}{r_b+h_b-c\cdot(r_b+h_b)}}+\gamma_z\right)\cdot\frac{1}{\gamma_{bm}}\right) \text{ with } \gamma_{zl}\le\gamma_z\le\gamma_{zp};$$

$$t_{kl} = \frac{1}{2\cdot\pi\cdot n_b}\cdot\arcsin\left(-\frac{L}{2\cdot H}\right) \text{ with } \gamma_{zlm}\le\gamma_z<\gamma_{zl}; \tag{14}$$

$$t_{kp} = \frac{1}{2\cdot\pi\cdot n_b}\cdot\arcsin\left(\left(\sqrt{\frac{2\cdot(h_z-h_b)}{r_b+h_b-c\cdot(r_b+h_b)}}+\gamma_z\right)\cdot\frac{1}{\gamma_{bm}}\right) \text{ with } \gamma_{zl}\le\gamma_z\le\gamma_{zp};$$

$$t_{kp} = \frac{1}{2\cdot\pi\cdot n_b}\cdot\arcsin\left(\frac{L}{2\cdot H}\right) \text{ with } \gamma_{zp}\le\gamma_z<\gamma_{zpm}.$$

The values of the items $\gamma_{zl}, \gamma_{zp}, \gamma_{zlm}$ *u* γ_{zpm} in (14) are defined by the equations (15).

$$\gamma_{zl} = \sqrt{\frac{2\cdot(h_z-h_b)}{(r_b+h_b)\cdot(1-c)}} - \gamma_{bm};$$

$$\gamma_{zlm} = \frac{\Delta L}{r_b+h_b} - \gamma_{bm}; \tag{15}$$

$$\gamma_{zp} = \gamma_{bm} - \sqrt{\frac{2\cdot(h_z-h_b)}{(r_b+h_b)\cdot(1-c)}}$$

$$\gamma_{zlm} = \gamma_{bm} - \frac{\Delta L}{r_b + h_b}$$

Similarly, we find the moment of contact of the grain and the workpiece with the maximum depth of cut:

$$t_{bm} = \frac{1}{2 \cdot \pi \cdot n_b} \cdot \arcsin\left(\frac{\gamma_z}{\gamma_{bm}}\right). \tag{16}$$

Integrating equation (13) in the limits from the start of contact with the surface to be machined t_{kl} up to time period $t < t_{kp}$, we determine the one-time energy expended by a single grain performing a complete cut during the time of continuous cutting $t - t_{ki}$:

$$A_{zp} = 34 \cdot \tau_s \cdot (1 + 1{,}5 \cdot f \cdot (1 + \xi)) \cdot \sqrt{v_d^2 + v_{bo}^2} \times$$
$$\times \int_{t_{kl}}^{tp} \left(h_z - h_b - \frac{1}{2} \cdot (r_b + h_b) \cdot (sin(2 \cdot \pi \cdot n_b \cdot t) \cdot \gamma_{bm} - \gamma_z)^2 (1 - c)\right)^2 \times$$
$$\times \sqrt{1 - \frac{v_{bo}^2}{v_d^2 + v_{bo}^2} \cdot sin^2(2 \cdot \pi \cdot n_b \cdot t)} \cdot dt.$$

By substituting:

$$\varepsilon = \frac{t}{t_{bm}}, \text{ where } dt = t_{bm} \cdot d\varepsilon,$$

we take the integral expression of the given equation to a dimensionless form:

$$A_{zp} = 34 \cdot \tau_s \cdot (1 + 1{,}5 \cdot f \cdot (1 + \xi)) \cdot \sqrt{v_d^2 + v_{bo}^2} \cdot (h_z - h_b)^2 \cdot t_{bm} \cdot J_{zp} \tag{17}$$

where is

$$J_{zp}=\int_{\frac{t_{kl}}{t_{bm}}}^{\frac{tp}{t_{bm}}}\left(1-\frac{r_b-h_b}{2\cdot(h_z-h_b)}\cdot(\sin(2\cdot\pi\cdot n_b\cdot t_{bm}\cdot\varepsilon)\cdot\gamma_{bm}-\gamma_z)^2(1-c)\right)^2\times$$

$$\times\sqrt{1-\frac{v_{bo}^2}{v_d^2+v_{bo}^2}\cdot\sin^2(2\cdot\pi\cdot n_b\cdot t_{bm}\cdot\varepsilon)}\cdot d\varepsilon.$$

As in (17), we find the one-time energy of cutting with an abrasive grain, which performs an incomplete cut and drops out of the working surface of the tool at the point in time t at the depth $a_z=a_{kp}$:

$$A_{zt}=34\cdot\tau_s\cdot(1+1{,}5\cdot f\cdot(1+\xi))\cdot\sqrt{v_d^2+v_{bo}^2}\cdot(h_z-h_b)^2\cdot t_{bm}\cdot J_{zt},\tag{18}$$

where is

$$J_{zt}=\int_{\frac{t_{kl}}{t_{bm}}}^{\frac{t}{t_{bm}}}\left(1-\frac{1}{2}\cdot\frac{r_b+h_b}{h_z-h_b}\cdot(sin(2\cdot\pi\cdot n_b\cdot t_{bm}\cdot\varepsilon)\cdot\gamma_{bm}-\gamma_z)^2(1-c)\right)^2\times$$

$$\times\sqrt{1-\frac{v_{bo}^2}{v_d^2+v_{bo}^2}\cdot sin^2(2\cdot\pi\cdot n_b\cdot t_{bm}\cdot\varepsilon)}\cdot d\varepsilon.$$

We define the following:

$$x=\frac{a_z}{h_z-h_b}=1-\frac{(r_b+h_b)\cdot(1-c)}{2\cdot(h_z-h_b)}\cdot(sin(2\cdot\pi\cdot n_b\cdot t)\cdot\gamma_{bm}-\gamma_z)^2\tag{19}$$

From this, it follows

$$t=\frac{1}{2\cdot\pi\cdot n_b}\cdot arcsin\left(\left(\pm\sqrt{\frac{2\cdot(h_z-h_b)\cdot(1-x)}{r_b+h_b-c\cdot(r_b+h_b)}}+\gamma_z\right)\cdot\frac{1}{\gamma_{bm}}\right).$$

Differentiating this equation, we get:

$$dt=\frac{1}{2\cdot\pi\cdot n_b\cdot\gamma_{bm}}\cdot\sqrt{\frac{2\cdot(h_z-h_b)}{(r_b+h_b)\cdot(1-c)}}\cdot\frac{1}{\sqrt{1-x\cdot}}\times$$

$$\times\frac{dx}{\sqrt{1-\frac{1}{\gamma_{bm}^2}\cdot\left(\gamma_z-\sqrt{\frac{2\cdot(h_z-h_b-x)}{(r_b+h_b)\cdot(1-c)}}\right)^2}}.$$

and the new limits of integration are:

$$x_{min}=0\ npu\ \gamma_{zl}\leq\gamma_z\leq\gamma_{zp};$$

$$x_{min}=1-\frac{(r_b+h_b)\cdot(1-c)}{2\cdot(h_z-h_b)}\cdot\left(\frac{L\cdot\gamma_{bm}}{2\cdot A}+\gamma_z\right)^2 \text{ with } \gamma_{zlm}\leq\gamma_z<\gamma_{zl};$$

$$x_t=\frac{a_z}{a_{zm}};$$

$$x_{max}=0 \text{ with } \gamma_{zl}\leq\gamma_z\leq\gamma_{zp};$$

$$x_{max}=1-\frac{(r_b+h_b)\cdot(1-c)}{2\cdot(h_z-h_b)}\cdot\left(-\frac{L\cdot\gamma_{bm}}{2\cdot A}+\gamma_z\right)^2 \text{ with } \gamma_{zpm}\leq\gamma_z<\gamma_{zp}.$$

Expressing equation (16) through a new variable, we determine:

$$A_{zx}=\frac{34\cdot\tau_s\cdot(1+1{,}5\cdot f\cdot(1+\xi))\cdot\sqrt{v_d^2+v_{bo}^2}\cdot(hz-hb)^{2,5}}{\sqrt{2}\cdot\pi\cdot n_b\cdot\sqrt{(r_b+h_b)\cdot(1-c))}\cdot\gamma_{bm}}\cdot J_{at} \tag{20}$$

,

where J_a is a dimensionless coefficient with $a_z \leq a_{kp}$

$$J_{zx} = \int_{x_{min}}^{x_{kp}} \frac{x^2}{\sqrt{(1-x)}} \cdot \sqrt{\frac{1 - \frac{v_{bo}^2}{\gamma_{bm}^2 \cdot \left(v_d^2 + v_{bo}^2\right)} \left(\gamma_z - \sqrt{\frac{2 \cdot (h_z - h_b) \cdot (1-x)}{(r_b + h_b) \cdot (1-c)}}\right)^2}{\left[1 - \frac{1}{\gamma_{bm}^2} \left(\gamma_z - \sqrt{\frac{2 \cdot (h_z - h_b) \cdot (1-x)}{(r_b + h_b) \cdot (1-c)}}\right)^2\right]}} \cdot dx. \quad (21)$$

The second factor in equation (21) characterizes the influence of the variable speed of travel of the abrasive stone along the surface to be machined on the cutting energy. But as the speed of travel of the stone during superfinishing is significantly lower than the peripheral speed of the workpiece, the value of this factor is close to one.

For practical calculations, the influence of the relation v_b / v_d in (21) can be neglected and the integrand simplified.

In order to analyse the effect of the position of the stone along the surface line of the workpiece on the cutting energy, it is important to express equation (13) through the relative ordinate of the cross-section of the workpiece. For this purpose, we introduce the definition:

$$u = z / H$$

as

$$z_b = H \cdot sin(2 \cdot \pi \cdot n_b \cdot t),$$ then

$$t = \frac{1}{2 \cdot \pi \cdot n_b} \cdot arcsin\, u ,$$

and

$$dt = \frac{1}{2 \cdot \pi \cdot n_b} \cdot \frac{du}{\sqrt{1-u^2}} .$$

Then we obtain:

$$dA = \frac{34 \cdot \tau_s \cdot (1 + 1{,}5 \cdot f \cdot (1 + \xi))}{2 \cdot \pi \cdot n_b} \cdot \sqrt{v_d^2 + v_{bo}^2} \cdot (h_z - h_b)^2 \times$$

$$\times \left(1 - \frac{(r_b + h_b) \cdot (1 - c)}{2 \cdot (h_z - h_b)} \cdot (u \cdot \gamma_{bm} - \gamma_z)^2 (1 - c) \right)^2 \times \qquad (22)$$

$$\times \sqrt{1 - \frac{v_{bo}^2}{v_d^2 + v_{bo}^2} \cdot u^2} \cdot \frac{du}{\sqrt{1 - u^2}} \text{ with } u_{\min} \le u \le u_{\max},$$

whereis

$$u_{\min} = \left(-\sqrt{\frac{2 \cdot (h_z - h_b)}{(r_b + h_b) \cdot (1 - c)}} + \gamma_z \right) \cdot \frac{1}{\gamma_{bm}} \text{ with } \gamma_{zl} \le \gamma_z \le \gamma_{zp};$$

$$u_{\min} = -0.5 \cdot \frac{L}{H} \text{ with } \gamma_{zlm} < \gamma_z \le \gamma_{zl}; \qquad (23)$$

$$u_{\max} = \left(\sqrt{\frac{2 \cdot (h_z - h_b)}{(r_b + h_b) \cdot (1 - c)}} + \gamma_z \right) \cdot \frac{1}{\gamma_{bm}} \text{ with } \gamma_{zl} \le \gamma_z \le \gamma_{zp};$$

$$u_{\max} = 0.5 \cdot \frac{L}{H} \text{ with } \gamma_{zp} < \gamma_z \le \gamma_{zpm}.$$

Integrating the expression (22) with respect to the variable u, we determine:

$$A_{zu} = \frac{34 \cdot \tau_s \cdot (1 + 1{,}5 \cdot f \cdot (1 + \xi))}{2 \cdot \pi \cdot n_b} \cdot \sqrt{v_d^2 + v_{bo}^2} \cdot (h_z - h_b)^2 \cdot J_u, \qquad (24)$$

where is

$$J_{zu} = \int_{u_{min}}^{u} \left(1 - \frac{1}{2} \cdot \frac{r_b + h_b}{h_z - h_b} \cdot (u \cdot \gamma_{bm} - \gamma_z)^2 (1-c) \right)^2 \times$$

$$\times \frac{1}{\sqrt{1-u^2}} \sqrt{1 - \frac{v_{bo}^2}{v_d^2 + v_{bo}^2} \cdot u^2} \cdot du.$$

It is easy to show that the energy of a single cut reflects the volume of the single cut. The volume of the chip removed by a single abrasive grain during the period of time dt, equals:

$$dV_z = S_z \cdot \vec{v}_r \cdot dt \tag{25}$$

where $S(\tau)$ is the sectional area of the groove left by the grain at the point in time τ.

In [1] it is shown that the parameters of the cutting part of the grain are not strictly defined. In many ways, they are dependent on the depth of penetration of the grain into the surface to be machined and the material of the abrasive grains. The deeper the grain penetrates into the surface to be machined, the higher is the number of irregularities of the abrasive grain that constitute the profile of its cutting part. And for the formation of the cutting part of the grain larger asperities start to gain significance. Therefore, with an increasing depth of penetration of the grain into the surface of the workpiece, the radius of the cutting part of the grain is automatically increased. For example, for aluminium oxide grains, the radius of the cutting grains is:

$$\rho_z = 6{,}83 \cdot \frac{a^{1,14}}{d_z^{\,0,14}},$$

where d_z is the diameter of the abrasive grain in mm.

With the increase in the radius of the cutting part, the grain becomes more solid, and the grain begins to cut thicker chips. Thus, a natural self-regulation process starts to manifest in the grain, in which the grain can cut both very thin chips at low depths of cut and considerably thicker chips at significant depths of penetration of the grain into the surface to be machined. Such an effect cannot be achieved for artificial cutting tools.

Naturally, with the increase in radius of the cutting grains and increasing depths of penetration of the grain into the surface to be machined, also the width of the grooves scratched by the grain increases:

$$b_z = 2{,}83 \cdot \frac{a_z^{\ 1{,}07}}{d_z^{\ 0{,}07}} \approx kb \cdot a_z .$$

In fine-grained abrasive tools with grain diameters that vary from 0.007 mm to 0.016 mm, the value of kb varies from 4.0 to 3.8, i.e., less than 5%. Therefore, we take

$$b_z = 3{,}9 \cdot a_z . \qquad (26)$$

Thus, if we present the profile of a single abrasive grain in the form of a parabola, then:

$$S_z = \frac{2}{3} \cdot a_z \cdot b_z = 2{,}6 \cdot a_z^2 , \qquad (27)$$

where a_z and b_z are the depth and width of the groove produced by the grain on the surface of the workpiece at the point in time t , mm.

Substituting the value of the scratch area (27) in equation (25) we find:

$$dV_z = 2{,}6 \cdot a_z^2 \cdot \vec{v}_r \cdot dt .$$

Solving for a_z and v_r in this term and integrating over time, we determine:

$$V_z = 2{,}6 \cdot \int_{t_{ki}}^{t} \left(h_z - h_b - \frac{(r_b + h_b)}{2} \cdot (\sin(2 \cdot \pi \cdot n_b \cdot t) \cdot \gamma_{bm} - \gamma_z)^2 (1 - c) \right)^2 \times$$
$$\times \sqrt{v_d^2 + v_{bo}^2 \cdot \cos^2(2 \cdot \pi \cdot n_b \cdot t)} \cdot dt = 2{,}6 \cdot (h_z - h_b)^2 \cdot \sqrt{v_d^2 + v_{bo}^2} \cdot J_{At}. \quad (28)$$

From a comparison of equations (17) and (28) follows that the volume of a single cut is similar to the energy of a single cut:

$$V_z = K_v \cdot A_z, \quad (29)$$

where K_v is the dimensional similarity coefficient, 1/MPa, equal to

$$K_v = \frac{0{,}0765}{\tau_s \cdot (1 + 1{,}5 \cdot f \cdot (1 + \xi))}.$$

Thus, the energy of a single cut represents a parameter such as the volume of a single cut, but in relation to it is a more general parameter, as it also reflects the properties of the material to be machined – shear stress and cutting conditions – the coefficient of friction of the grain and the surface to be machined f and the chip compression ξ.

The average area of the cut is determined by integrating the current area of the cut (27) with respect to time and dividing the result by the time of formation of the scratch:

$$S_{zs} = \frac{2}{3 \cdot (t - t_{ki})} \cdot \int_{t_{ki}}^{t} a_z \cdot b_z \cdot dt$$

Using the above relation between the width and the depth of cut (26), and solving for the value of the depth of cut (27), we determine:

$$S_{zs} = \frac{2{,}6}{t - t_{ki}} \cdot \int_{t_{ki}}^{t} \left(h_z - h_b - \frac{1}{2} \cdot (r_b + h_b) \cdot (\sin(2 \cdot \pi \cdot n_b \cdot t) \cdot \gamma_{bm} - \gamma_z)^2 (1 - c) \right)^2 \cdot dt =$$
$$= \frac{2{,}6}{t - t_{ki}} \cdot (h_z - h_b)^2 \cdot J_S,$$

where is

$$J_S = \int_{t_{kl}}^{t} \left(1 - \frac{r_b + h_b}{2 \cdot (h_z - h_b)} \cdot (\sin(2 \cdot \pi \cdot n_b \cdot t) \cdot \gamma_{bm} - \gamma_z)^2 (1 - c) \right)^2 \cdot dt,$$

From a practical point of view, it is more convenient to express the area of the cut through the depth of cut. Using the same substitution as in equation (17), we determine after the transformations:

$$S_s = 2{,}6 \cdot (h_z - h_b)^2 \cdot J_S, \tag{30}$$

where is

$$z_{kp} = \left(\sqrt{\frac{2 \cdot (h_z - h_b)}{r_b + h_b - c \cdot (r_b + h_b)}} + \gamma_z \right) \cdot \frac{H}{\gamma_{bm}} \quad \text{with } \gamma_{zl} \le \gamma_z \le \gamma_{zp};$$

$$J_S = \int_{x_{min}}^{x_{kp}} \frac{x^2}{\sqrt{(1-x)}} \cdot \sqrt{\frac{1}{\left[1 - \frac{1}{\gamma_{bm}^2} \left(\gamma_z - \sqrt{\frac{2 \cdot (h_z - h_b) \cdot (1-x)}{(r_b + h_b) \cdot (1-c)}} \right)^2 \right]}} \cdot dx.$$

From a comparison of equations (18) and (30), we find that:

$$S_{zs} = \frac{0{,}0765 \cdot J_S \cdot A_t}{\tau_s \cdot (1 + 1{,}5 \cdot f \cdot (1 + \xi)) \cdot (t - t_{ki}) \cdot \sqrt{v_d^2 + v_{bo}^2} \cdot J_{At}} \tag{31}$$

We find the average depth of penetration of a grain into the surface to be machined. Typically, the abrasive grain penetrates into the surface of the workpiece gradually from 0 up to its maximum value. The maximum value depends on the fact that the grain drops out of the bond under the influence of the cutting forces or performs a complete cut.

We assume the abrasive grain is at an angle γ_z from the plane of symmetry of the tool, penetrates into the surface to be machined with a maximum depth $a_{zm} = h_z - h_b$ and makes a complete cut. In this case, the depth of penetration of the grain into the surface of the workpiece is determined by equation (14). The start of the contact of the top of the grain with the surface to be machined (touchdown) corresponds to the position of the tool, in which the coordinate of its point of contact is z_{kl} (17), and the maximum depth of penetration of the grain in the surface to be machined corresponds to the coordinate $z_{bv} = z_{bm}$. Then the mean depth of penetration of the grain into the surface to be machined is:

$$a_i = \frac{1}{z_{bv} - z_{kl}} \cdot \int_{z_{ki}}^{z_{bv}} [h_z - h_b - \frac{1}{2} \cdot (r_b + h_b) \cdot (\frac{z_b}{H} \cdot \gamma_{bm} - \gamma_z)^2 (1-c)] \cdot dz_b \quad (32)$$

.

After the transformations (32), as was to be expected, we get:

$$a_i = \frac{2}{3} \cdot a_{zm} \quad . \quad (33)$$

If the grain performs an incomplete cut and drops out of the bond at $a_z = a_{kp}$, the upper limits of integration are:

$$z_{bv} = \left(-\sqrt{\frac{2 \cdot (h_z - h_b) \cdot (1 - a_{kp})}{r_b + h_b - c \cdot (r_b + h_b)}} + \gamma_z \right) \cdot \frac{H}{\gamma_{bm}} \qquad \text{with } \gamma_{zl} \le \gamma_z \le \gamma_{zp}; \quad (34)$$

.

Then, we obtain the average depth of scratch from equation (17):

$$a_{zs} = 0{,}17 \cdot \sqrt{\frac{J_S \cdot A_t}{\tau_s \cdot (1 + 1{,}5 \cdot f \cdot (1 + \xi)) \cdot (t - t_{ki}) \cdot \sqrt{v_d^2 + v_{bo}^2} \cdot J_{At}}} \quad (35)$$

.

It is easy to show that for the process of superfinishing, the relation $k_a = \sqrt{J_S / J_{At}}$ is close to one.

It is interesting to look at the dependency of the energy of cutting with an abrasive grain on the length of the cutting path. This all the more as the length of the cutting path is an independent energy parameter.

The elementary length of the cutting path of an abrasive grain is:

$$dl_{zt} = \vec{v}_r \cdot dt$$

Then, the length of the cutting path is determined by integrating this equation:

$$\begin{aligned} l_{zt} &= \int_{t_{kl}}^{t} \sqrt{v_{bo}^2 \cdot cos^2(2 \cdot \pi \cdot n_b \cdot t) + v_d^2} \cdot dt = \\ &= \sqrt{v_d^2 + v_{bo}^2} \cdot \int_{t_{ki}}^{t} \sqrt{1 - \frac{v_{bo}^2}{v_d^2 + v_{bo}^2} \cdot sin^2(2 \cdot \pi \cdot n_b \cdot t)} \cdot dt. \end{aligned} \tag{36}$$

By expanding the integrand (36) in the series and limiting this to two terms of the series, we find:

$$l_{zt} = \sqrt{v_d^2 + v_{bo}^2} \cdot \left(1 - \frac{v_{bo}^2}{4 \cdot (v_d^2 + v_{bo}^2)} \right) \cdot (t - t_{ki}).$$

The error in the given formula with $v_{bo} / v_d \leq 0{,}5$ is less than 1%. Substituting in this equation the value of time, expressed in the expression

$$\tau_m = \frac{2 \cdot P_{zk} \cdot h_y}{d_h^3 \cdot J_{i\tau}} \cdot \left(1 + \frac{2h_y}{d_h} \right) \cdot \sqrt[3]{\frac{(v_s + v_p)^2}{v_s^2}} \cdot \left(1 - \frac{2 \cdot h_y}{d_h \cdot \left(1 - \frac{2h_s}{d_h} \right)} \right)$$

through the depth of cut a_z, we get:

$$l_{zt} = \sqrt{v_d^2 + v_{bo}^2} \cdot \left(1 - \frac{v_{bo}^2}{4 \cdot (v_d^2 + v_{bo}^2)} \right) \cdot (\arcsin(\frac{1}{\gamma_{bm}} \cdot \left(\gamma_z - \sqrt{\frac{2 \cdot a_m - a_z}{(r_b + h_b) \cdot (1 - c)}} \right) -$$
$$- \arcsin(\frac{1}{\gamma_{bm}} \cdot \left(\gamma_z - \sqrt{\frac{2 \cdot a_m}{(r_b + h_b) \cdot (1 - c)}} \right)), \quad (37)$$

where $a_m = h_z - h_b$ is the maximum possible depth of the scratch with the given grain.

The vast majority of abrasive grains performs complete cuts, i.e., they touch the surface to be machined and get out of contact with the surface at a depth of cut close to zero. As follows from equation (37), the length of a complete cut is equal to:

$$l_z = \frac{2 \cdot H}{\gamma_{bm}} \cdot \sqrt{\frac{2 \cdot a_m}{(r_b + h_b) \cdot (1 - c)}} . \quad (38)$$

From equation (38) follows that the lengths of the complete cuts are proportional to the energy of the cuts.

CONCLUSION

The construction of a model of the superfinishing process is based on the following conditions:

(1) The number of energy parameters of the superfinishing process used should be sufficient for the necessary characteristics of the finishing process, such as performance, quality of finishing, amounts of stock removal, etc. The most universal criteria of this type include the one-time cutting energy and the specific force of superfinishing. The one-time cutting energy, on the one hand, is determined by the area of the single cuts, and, consequently, the depth of the residual scratches on the surface of the workpiece and the roughness of the obtained surface. On the other hand, this criterion depends on the length of the single cuts, and thus, together with the depth of cut, characterizes the

intensity of stock removal. Usually, with an increasing rate of stock removal, the roughness of the surface increases. But this is not always true, as an increase in cutting speed leads to an increase in one-time cutting energy at a constant depth of cut, and, therefore, it is possible to find conditions to increase the processing capacity without the need to increase the roughness of the surface to be machined.

In contrast to the one-time cutting energy, the power expended during finishing characterizes the total effect of all abrasive grains that are on the surface of contact with the workpiece on the surface to be machined. On the one hand, it depends on the one-time cutting energy, and therefore reflects the amount of roughness of the finished surface of the workpiece and the rate of stock removal. However, this parameter provides very important additional information about the process under study – namely, it reflects the impact of the abrasive grains participating in the contact and the size of the actual contact area.

Thus, the analysis of only these two energy parameters provides sufficient information about the mechanism of the superfinishing process sequence, and, consequently, allows for the determination of the rational conditions of the implementation and improvement of the process.

(2) The energy criteria of the superfinishing process should be universal and should allow for a comparison between the different methods and conditions for its implementation. This provides the opportunity not only to build a particular finishing process in a rational way through the optimisation of the conditions for its implementation. But it also offers the opportunity of an efficient search for the most rational ways to improve the process and allows creating new ways of processing and designing new superfinishing technologies. Such universal criteria include, again, the one-time cutting energy and the expended power. Based on whatever scheme the finishing process is performed, the one-time cutting energy is always determined by the one-time depth of cut and the cutting speed, and the expended power depends on the one-time cutting energy and the number of active abrasive grains that participate in the cutting operation within a given unit of time.

(3) The energy parameters of the superfinishing process must be mathematically related to the basic indicators of the process. It is of particular importance to provide the link of these parameters to such

important indicators as the metal removal during the finishing process and the roughness of the finished surface.

Processing by abrasion is characterized by a variety of random factors that determine the given process. These include the random shape and size of the abrasive grains, the random nature of their location on the working surface of the tool, both in depth and along the working surface, the random nature of the arrangement of bond posts in the tool, the random values of parameters characterizing the strength of retention of the grains in the bond, the random distribution of the pore volume on the working surface of the tool, the random depth of cuts, the random number of grains concurrently participating in cutting, etc. Therefore, deterministic models of abrasive finishing processes, as offered by most researchers, poorly reflect the actual process in this regard to date, and it became necessary to develop a coherent theory of abrasive finishing, which takes into account the stochastic nature of this complex multifactorial process.

(4) The energy of a single cut is a generalized parameter of the cut, as it takes into account both the cutting force, and depth, and length, and volume of the single cut. But each of these parameters characterizes the cut from one point of view, and the parameter of the energy of a single cut characterizes the nature of a cut in general.

REFERENCES

[1] Korolev, A.V. (1975). The study of the formation of the tool and workpiece surfaces with abrasive machining.Saratov: Saratov State Technical University, S192.

[2] Redko, S.G. (1962). Processes of heat development when grinding metals.Saratov: Saratov University Press, 299 p.

[3] Korczak, S.N. (1974). Performance of the grinding process of steel parts.Moscow: Mechanical engineering, 280 p.

In: Recent Advances in Abrasives Research
Editor: Dirk Bähre
ISBN: 978-1-62808-566-2

Chapter 5

The Effects of Consumables on Material Removal and Scratch Propensity in Chemical Mechanical Planarization

Abhijit Chandra,[1,2,*] Ashraf Bastawros,[2,1] Kuan-Chuen Wu,[1] Pavan Karra[1] and Tianyu Yu[2]

[1]Department of Mechanical Engineering,
Iowa State University, Ames, Iowa, US
[2]Department of Aerospace Engineering and Engineering Mechanics, Iowa State University, Ames, Iowa, US

Abstract

Chemical Mechanical Planarization (CMP) is a significant processing step in the manufacture of integrated circuits (IC) with feature sizes below 250 nm. In addition to local and global planarity of the workpiece resulting from a pressure differential activated material removal rate (MRR), defectivity of the workpiece and yield of the process are important benchmarks of performance. In particular, the scratch distribution profile of the workpiece after CMP is a notable aspect of defectivity that determines product quality, and MRR is a major metric of process efficiency.

---

* Corresponding author: Abhijit Chandra. E-mail: achandra@iastate.edu.

Properties of the consumables used in CMP, such as polishing pad topography, slurry solvent chemistry, and abrasive particle size, influence the scratch formation and the material removal process. This chapter examines the role of consumables in defectivity and yield of a CMP process by presenting analytical models for the time evolution of scratch generation propensity and MRR that are integrated into a multi-scale model and implemented in simulation software called "CMPSim" whose predictions of nano-scale scratch profiles and wafer-scale MRR compare well with empirical observations. The computational platform produces results that support experimental findings on the effects of certain key choices for consumables on the defectivity and yield in CMP.

Keywords: Chemical mechanical planarization, material removal, scratch propensity, surface defectivity, computer simulation

1. Introduction

Chemical Mechanical Planarization (CMP) is one of the most promising and prevalent technologies in metallization for ultra-large-scale integration because it can provide local and global planarization effectively [41, 9]. CMP is a preferred process for achieving high-precision surface flatness in several industries, including integrated circuit fabrication which demands smooth, planar semiconductor wafers and surfaces (conductor, semiconductor, and nonconductor) for traditional designs, and smooth, planar layers and films for unconventional, multilayer designs [7, 9]. Because CMP can be applied to a wide array of workpiece materials besides silicon and copper, including germanium, polysilicon, photoresist, silicon dioxide ("oxide"), silicon nitride, gallium nitride, zinc oxide, aluminum, tungsten, glass, polymers, and carbon nanotubes [41, 32], its applications and usefulness span multiple industries. A thorough understanding of the theoretical principles and the physical effects of consumable components in CMP can lead to improvements in product quality and process efficiency that may cross many boundaries.

There are many varieties of CMP configurations that satisfy diverse requirements, objectives, and constraints. The breadth of CMP equipment designs shows that the technology offers ample room for exploration, experimentation, and improvement. This chapter uses a representative process design to convey the universally applicable principles and strategies behind CMP. Further information about CMP equipment and process design is available from other sources [27, 41, 33].

The fundamental principle of operation in CMP is to leverage the dual, interacting phenomena of mechanical abrasion and chemical reaction through the use of consumable materials in order to polish a workpiece to a high level of surface smoothness, planarity, and finish quality. Smoothness refers to the degree of absence of local variation in surface height, i.e. the absence of surface roughness. Planarity refers to the degree of absence of global variation in surface height across the entire workpiece, i.e. the absence of plateaus and valleys. Finish quality refers to the degree of absence of defects such as scratches, cracks, and workpiece delamination. Different types of scratches on copper, shown in Figure 1, may be symptoms of different causal mechanisms. In IC manufacturing, even a scratch of nano-scale dimensions (depth, width, or length), such as the one appearing in Figure 1d, can cause immediate or eventual failure of the product [7]. Locally (e.g. at the feature scale), CMP effectively reduces variations in step heights, but globally (e.g. at the wafer scale), CMP may yield non-uniform planarity and MRR [47, 14]. One of the most significant challenges in CMP is to minimize problematic scratching, which affects product quality, while maximizing MRR, which affects process efficiency [25]. This chapter presents some models and recommendations on how to meet the challenge through strategic choices for CMP consumables.

Mechanical abrasion is performed by a polishing pad that can possess a variety of morphology, topography, and viscoelastic properties [33, 9]. Morphology can be described in terms of material microstructure; cellular and fibrous are two common categories of CMP pads used in research and industry. In the case of cellular pads, topography can be expressed in terms of attributes like porosity. Several viscoelastic properties are important because the pad material must be chosen in consideration of the workpiece material and the operational requirements. The pad's hardness, Young's modulus, and Poisson's ratio must be sufficient to abrade the workpiece when contact is made. Moreover, the pad's Critical Surface Tension needs to be high enough because hydrophilicity is required for proper wetting conditions, especially near the center region of the pad which tends to be away from the slurry inlet [33]. Furthermore, the pad's shear strength and tensile strength have to endure abrasion from micro-contact with the slurry particles, and to endure tearing from the forces encountered during CMP that tend to pull the pad apart. The viscoelastic properties of polyurethane are favorable according to the previous criteria, which is why it is a popular choice made available in many different forms (e.g. IC1000). Lastly, the polishing pad may have various thicknesses, grooves, shapes, and layers; treatment of these topics is available in other sources [33].

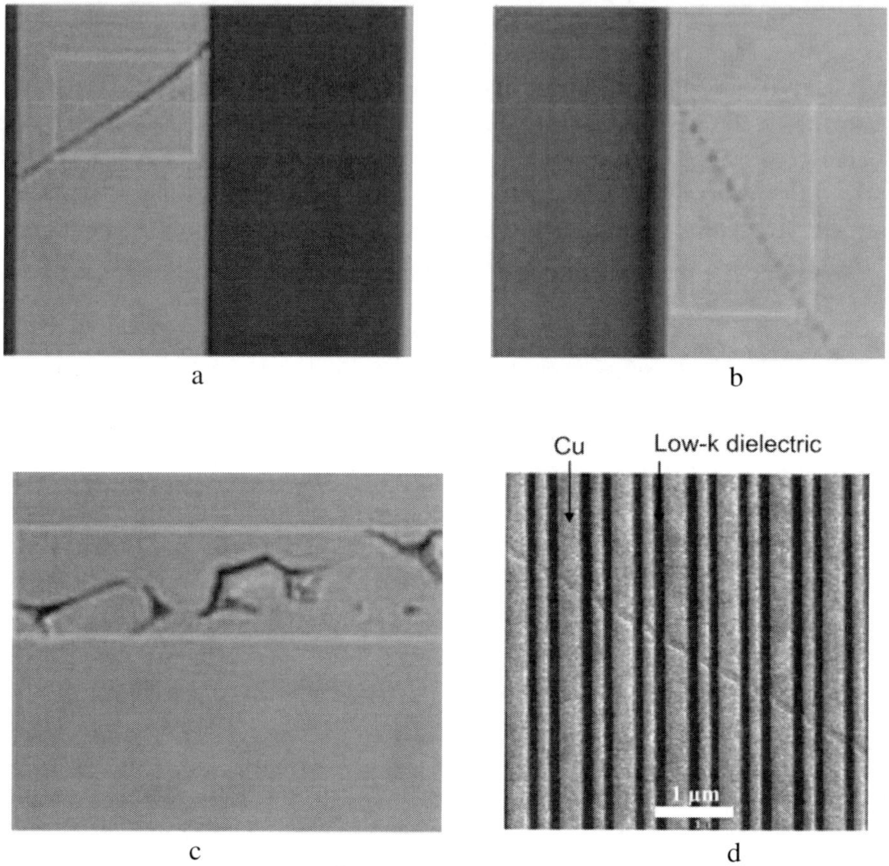

Figure 1. Different types of scratches on copper surface; (a) Razor scratch, adapted from [9]; (b) Chatter scratch, adapted from [9]; (c) Skipping scratch, adapted from [9]; (d) Nano-scratch on integrated circuit [12].

Chemical reaction is performed by a slurry compound that traditionally is a colloidal liquid suspension of solvents and abrasive particles with other agents such as pH buffers and adjusters, passivators, inhibitors, catalysts, stabilizers, and complexing agents, although variants such as particle-free slurries also exist [33, 9, 7]. The solvent is usually chosen to chemically attack (often, to oxidize) the workpiece material. For instance, an acid or hydrogen peroxide ("peroxide") can be chosen for a metal workpiece, though an alkaline solvent can also be used [36, 29]; each choice has advantages and disadvantages.

As another example, water can be the primary solvent for an oxide-layered workpiece. Other choices of solvent include KIO_3, $Fe(NO_3)_3$, and H_5IO_6.

Abrasive particles come in different sizes and formulations (silica SiO_2, alumina Al_2O_3, and ceria CeO_2 are commonly used), and their properties are chosen in the context of other operational requirements. For instance, each particle formulation has a unique iso-electric point that governs its agglomeration or coagulation behavior in bulk solution, which depends on the pH environment that is largely dictated by the solvent, so the choice of particle formulation should fit with the choice of solvent in consideration of factors such as likelihood of generating nano-scratches. Likewise, the concentrations of solvent and of abrasive particles should be chosen in a broader context that balances factors such as process efficiency (e.g. MRR) and product quality (e.g. defectivity). The slurry also performs other process-critical functions, such as pad-workpiece interfacial lubrication, heat transfer away from the polishing site, and workpiece debris transport away from the interface.

To perform mechanical abrasion and chemical reaction simultaneously in polishing a workpiece (e.g. a semiconductor wafer in the IC industry) with good surface smoothness, planarity, and finish quality, the CMP system diagrammed in Figure 2 may be used.

Before and after polishing, the carrier moves the workpiece toward and away from the pad by using various means, e.g. vacuum or negative gauge pressure on the backside of the workpiece. During polishing, the workpiece is kept in place underneath the carrier by various methods, e.g. a retainer ring surrounding the workpiece, and downward pressure is applied by the carrier in various ways, e.g. positive gauge pressure or piston actuation on the backside of the workpiece, so that there is mechanical contact with the pad.

Several types of rotary motion may occur in CMP. Typically, the carrier spins the workpiece about its axis, and the platen spins the pad about its axis. In some systems, the carrier may orbit around the platen's axis, while in other systems, the carrier may traverse linearly between the platen's center and perimeter. The directions and rates of rotation and revolution may be chosen according to different needs and strategies. The slurry is generally dispensed onto the pad so that the pad is properly wetted for simultaneous chemical and mechanical action on the workpiece, as illustrated in Figure 3.

As CMP proceeds, the pad gradually wears down; for example, as the pad's surface roughness decreases, the MRR declines [31, 25, 33, 38, 45], and as particles agglomerate and embed into the pad surface or get trapped between pad asperities, the probability of generating a scratch increases.

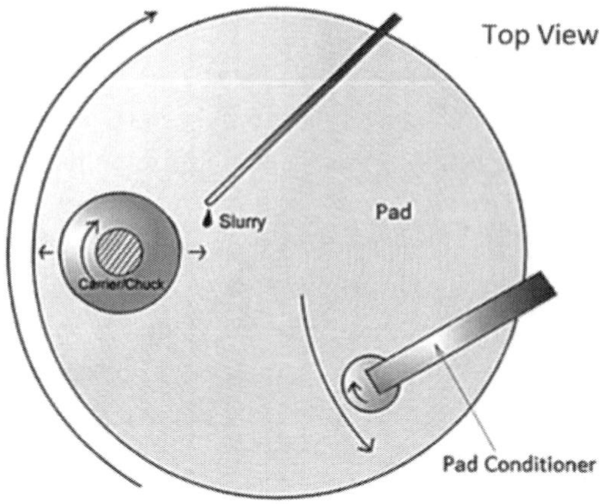

a

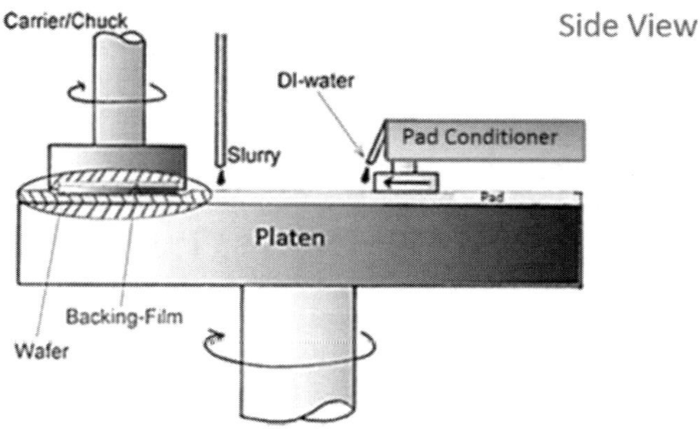

b

Figure 2. Schematic of representative CMP system; (a) Overhead view [32]; (b) Side view [32].

To counteract pad wear and its negative consequences, some CMP systems have a pad conditioner device (as depicted in Figure 2) that uses a hard abrasive like diamond grit to restore the porosity and roughness of the pad's surface.

Because of the multi-scale physics of its multiple components and their interactions [8], CMP offers latitude for control and optimization through two categories of variables. The first category of CMP variables comprises the parameters of the overall process, such as pressure, velocity, and temperature.

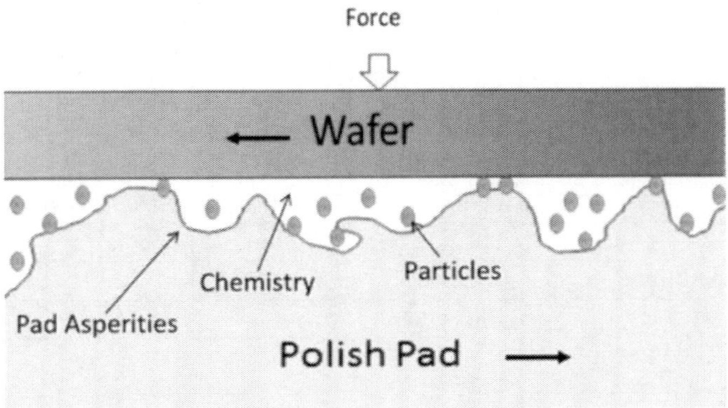

Figure 3. Simultaneous chemical and mechanical actions on wafer [32].

Pressure refers to the nominal downward force distributed over the carrier's area of contact with the workpiece; the actual pressure exerted by the workpiece onto the pad (or vice versa) generally has a very different value, for reasons which are discussed later in this chapter.

Velocity refers to some measure of the relative speed (perhaps in an average sense) between the workpiece and the pad, which depends on other velocities such as the carrier's rotational velocity and linear velocity (if it slides relative to the platen), the platen's rotational velocity, and the carrier's orbital velocity (if there is orbital motion). Temperature refers to the ambient temperature in which CMP is performed; there may well be thermal variations within the system and within the workpiece.

The second category of CMP variables comprises the many possible choices for the consumables; some of the key choices appear in Table 1 [41, 15, 9, 26]. As discussed earlier in this chapter, characteristics of the consumables are interrelated by physical laws and engineering configurations; therefore, choices for CMP consumables should be strategically made in order to achieve the objectives while satisfying the constraints.

Modeling of CMP has paramount importance in the understanding and improvement of the technology. The CMP process has been dissected and analyzed in many ways.

Models have been developed at different length scales from small to large, such as particle scale (nominally 1 nm to 100 nm), feature scale (about 100 nm to 10 μm), die scale (around 10 mm), and wafer scale (typically 10 mm to 150 mm), because different problems occur at different length scales [27].

Table 1. Some important characteristics of consumable materials in CMP

Pad	Slurry	Particle
Morphology, microstructure	Solvent chemistry and concentration	Formulation (hardness, iso-electric point)
Topography	pH	Size distribution
Hardness	Particle concentration	Shape
Young's modulus	Viscosity	
Poisson's ratio	Flow rate	
Critical surface tension, Liquid permeability		
Shear strength		
Tensile strength		

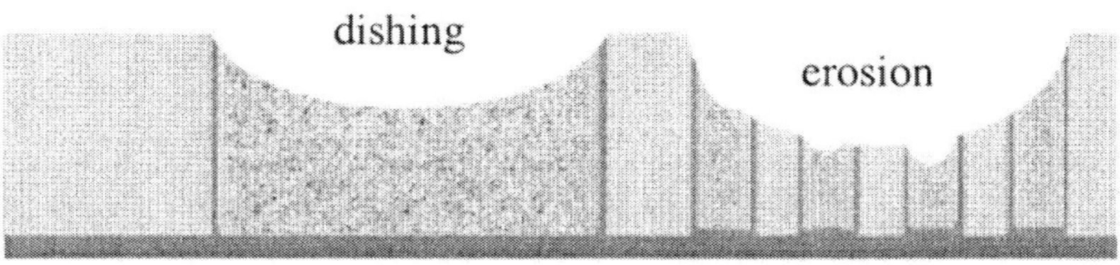

Figure 4. Examples of surface defects after CMP of copper [16].

For example, defects such as nano-scratches, nano-cracks, and metal residues may occur at the particle scale; metal dishing and insulator erosion (depicted in Figure 4), feature scale; non-uniform step height, die scale; and non-uniform MRR, wafer scale. Analytical models for MRR and scratch generation propensity are discussed later in this chapter. Few models fully capture the interactions between CMP process parameters at multiple length scales [25]. This chapter presents an experimentally supported model of the CMP process as it pertains to the MRR and to the formation of nano-scale scratches.

2. Development of Pad Model for Material Removal Rate

The polishing pad plays a significant, multifaceted role in CMP. The pad is involved in the removal of material from the workpiece and in the formation of scratches on the workpiece surface.

Also, the pad's function is intertwined with the slurry's function. It can be seen in Figure 2 and Figure 3 that when there is a non-zero relative velocity, the pad removes material when the asperities on its surface make mechanical contact with the workpiece, and also when the asperities mechanically "trap" abrasive particles that remove material upon contact with the workpiece. The scratch generation process is affected by the mechanism and mode in which the pad asperities "trap" the slurry particles, among other factors that are discussed later in this chapter. Also evident from Figure 2 and Figure 3 is that in CMP, the pad works in conjunction with the slurry to execute the combined chemical and mechanical actions on the workpiece. Specifically, the pad partially absorbs the slurry to some extent, and partially distributes the slurry across the interface with the workpiece. Assuming that the slurry is not reused by the CMP process, the pad dispenses clean, cool slurry containing new abrasive particles throughout the workpiece contact interface, and the pad carries away debris-laden slurry containing agglomerated particles and waste heat from the workpiece contact interface. Clearly, the pad and its physical properties are important in CMP, and accurate models of the pad are needed.

2.1. Pad Characterization

The morphology and topography of polishing pads both play a significant role in the CMP process. Two of the principal morphologies for CMP pads are porous (often termed "cellular") and fibrous. This chapter focuses on cellular pad morphology [3]. The IC1000 is a cellular polishing pad that exhibits representative characteristics [3, 5, 21, 44]. Its surface has numerous pores with an average diameter typically between 30 μm and 50 μm. The pores typically occupy about 35% of the pad's material volume and are separated by polyurethane membranes, which form asperities constituting roughness on the pad's surface. During a polishing process, it is the asperities that actually make contact with the workpiece surface. The real contact area is usually very small, about 1% or less of the nominal contact area between the workpiece and the pad [39]. Thus, the real contact pressure, which directly affects the MRR in CMP, is much higher than the nominal pressure. The pad topography can be characterized by a probability density function (PDF) of the pad asperities' heights; such an approximation aids the mathematical modeling process. Certain choices for the PDF, such as an exponential distribution, can reduce the computational burden when the numerical model is run [43].

If the pad asperity height PDF is a parametric distribution, it can be specified by statistical measures such as mean, standard deviation, skewness, and kurtosis.

Analysis of a representative CMP pad (e.g. IC1000) may provide some insight for the relationship between pad structure and MRR. The pad topography structure is shown in Figure 5. It has a cellular structure with partially connected pores whose average diameter is $D_p \approx 50$ µm. The locked-in porous structure, with pined nodal points and semi-attached cell membrane faces, result in higher initial stiffness and plateau stresses [18]. The pad surface density β is estimated by digital image processing as 35%, representing the average percentage of dense material in an arbitrary cross-section of the pad. The descriptors of the pad's surface roughness, relative to the pad's average cell diameter, may be measured with a surface profilometer at multiple length scales [3]. A long wavelength roughness λ ($\lambda \approx 15 - 20\ D_p$) and amplitude a ($a \approx 0.1 - 0.2\ D_p$) can be observed in Figure 6. It is worth noting that within the roughness of wavelength λ, there is a region of smaller wavelength λ_1 ($\lambda_1 \approx 2 - 4\ D_p$) wherein the cell membranes are continuous and slightly higher than the surrounding region; some examples are marked by the dashed circles in Figure 5.

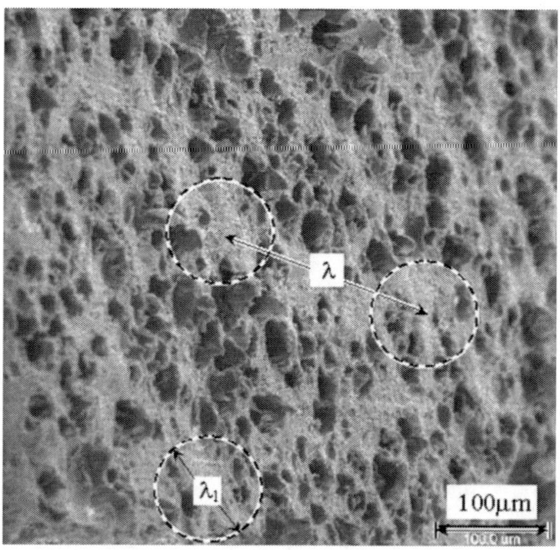

Figure 5. SEM image of IC1000 pad whose surface is tilted 60° off the viewing axis to show relative surface topography; marked areas (diameter $\lambda_1 \approx 100$ µm) are separated from one another by $\lambda \approx 800$ µm on average and slightly higher than surrounding areas [3].

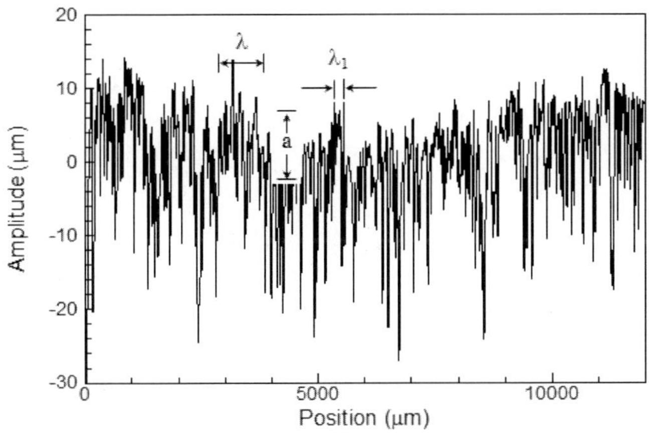

Figure 6. Pad surface topography with roughness wavelength $\lambda \approx 800$ μm, roughness amplitude $a \approx 12$ μm, and asperity wavelength $\lambda_l \approx 100$ μm [3].

The progressive contact of these long-range asperities makes controlling the pad's macroscopic mechanical response and the actual workpiece-pad contact more difficult. The roughness of the pad, at a scale comparable to its average pore diameter, affects the characteristics of force transmission from the externally applied pressure to the individual abrasive particles. To keep the model simple, the stochastic nature of the pad's surface roughness at this length scale, as well as the probabilistic distribution of the particles' size, are omitted.

To generate Figure 7a, the mechanical characteristics of a dry polishing pad are obtained from a compression test on a pad section having dimensions 5.64 mm x 5.55 mm x 1.02 mm without the soft back packing, in the Instron 8162 servo-mechanical testing frame operated under displacement control [3]. The corresponding compressive stress-strain curve is plotted in Figure 7a. The initial slope of the curve represents the initial Young's modulus E_o ($E_o \approx 180$ MPa) of the dry, porous pad.

The resulting curve indicates that the pad response is highly nonlinear, which is typical of cellular materials [18]. In addition, the initial response exhibits an extended contact range due to the pad's surface asperities and roughness.

To analyze the pad's initial response further, a 200 μm flat circular punch is pressed against the pad in a nano-indentor machine. The corresponding curve of force versus deflection, which is easily transformed into engineering stress versus strain via appropriate normalization, is plotted in Figure 7b.

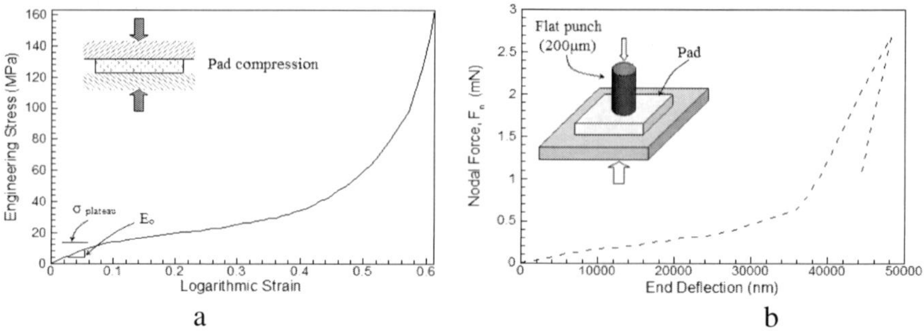

Figure 7. (a) Nonlinear compressive stress-strain curve for dry IC1000 pad without soft packing [3]; (b) Nonlinear stress-strain curve for IC1000-like dry pad (Thomas West P313) [3].

For a typical CMP process, the applied nominal pressure P_o is below or just slightly higher than the initial contact pressure, which puts the CMP process inside the nonlinear region of the pad's low-stiffness mechanical response that is caused by the asperities' contact with the workpiece. The time-dependent response of the pad material, the patterns of pad groove geometry, and the details of pad glazing and pad conditioning can be found in other sources [14, 32, 31, 49].

2.2. Contact Model

By itself, a static model of the polishing pad's morphology and topography is not sufficient for modeling a CMP process. A dynamic model of the pad's mechanical deformation response, when there is physical contact with the workpiece, is also needed. The pad-workpiece contact is a complex physical phenomenon that causes material removal, scratch formation, and pad wear, and it also involves the abrasive particles in the slurry, so a complete model of a CMP process must include the pad-particle interaction and the pad-workpiece interaction, which are addressed now; the particle-workpiece interaction is covered later in this chapter.

2.2.1. Material Removal Rate Model

Clearly, there are numerous ways to model a complex, multi-physics, multi-scale process such as CMP. The MRR is a prevalent output metric for validating CMP models due to its encompassment of most, if not all, of the relevant phenomena [27].

Among the numerous models of MRR, some build upon prior models, whereas others pursue a different train of thought. Many early models of MRR are deterministic in accounting for material removal by a single particle, aggregating particle-scale phenomena in order to predict wafer-scale quality control parameters like MRR.

Some later models employ probabilistic phenomena and mathematics to describe wafer-scale MRR at the particle scale [25]. For instance, there are models based on empirical measurements [37]; elastic solid-solid models and hydrodynamic models for mechanical contact between the abrasive particle(s) and the workpiece [27]; and models of material removal through cutting, plowing, and shearing at the intersection of nano-scale scratches [11]. Alternatively, MRR may be inferred or computed from a model of the wafer surface's time evolution [47]. This chapter presents a multi-scale, physical model for the MRR in CMP that is constructed as follows [3].

Greenwood et al. developed a model, depicted in Figure 8, for mechanical contact between a smooth, flat surface and a rough surface using Hertzian contact theory [2, 24, 6]. In the Greenwood contact model, the wafer is the workpiece. The model may be adapted to the CMP process with the goal of computer simulation if the wafer surface is assumed to be smooth and flat on the global scale, and the pad surface is assumed to be rough due to asperities having a height PDF $\varphi(Z_p)$ relative to the mean plane of the pad surface.

Moreover, the asperities are assumed to be uniformly spaced across the pad surface with area density η_s, and all asperities are assumed to have identical spherical tips of radius R_p [6].

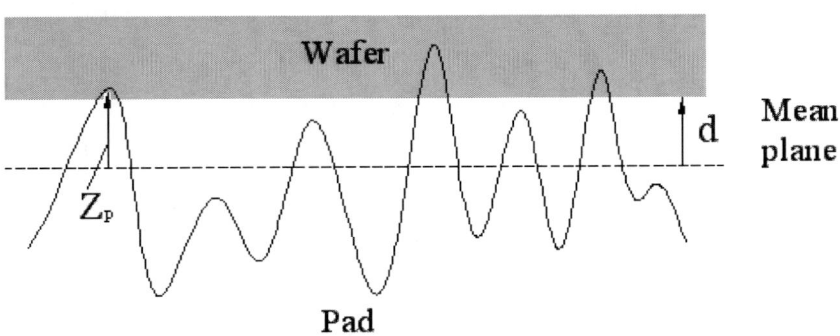

Figure 8. Illustration of Greenwood model [6].

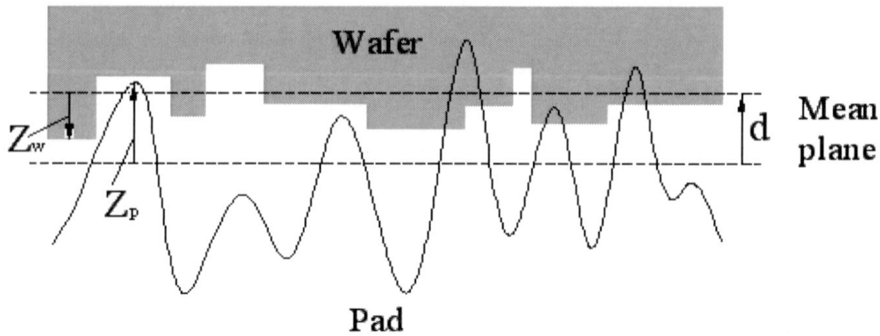

Figure 9. Schematic of contact model between rough pad and patterned workpiece [6].

The separation distance d in Figure 9 is defined as the distance between the mean plane of the pad surface, and the smooth wafer surface. According to Hertzian contact theory, an asperity of height $Z_p > d$ can carry a load

$$L(d) = \frac{4}{3} E^* \left(R^*\right)^{\frac{1}{2}} (Z_p - d)^{\frac{3}{2}} \tag{1}$$

where E^* is the equivalent Young's modulus for the asperity and R^* is the equivalent asperity tip radius as expressed in Equations (5) and (6), respectively. The contact area A between the asperity and the wafer is approximated by

$$A(d) = \pi R^* (Z_p - d) \tag{2}$$

The total load L_{total} and real contact area A_{re} for all pad asperities are equal to the mean load or mean contact area, respectively, of a single asperity multiplied by the total number of asperities:

$$L_{total}(d) = A_o \eta_s \int_{Z_p > d} \frac{4}{3} E^* \left(R^*\right)^{\frac{1}{2}} (Z_p - d)^{\frac{3}{2}} \varphi(Z_p) dZ_p \tag{3}$$

$$A_{re}(d) = A_o \eta_s \int_{Z_p > d} \pi R^* (Z_p - d) dZ_p \tag{4}$$

In Equations (3) and (4), A_o is the nominal contact area, and R^* is the equivalent asperity tip radius given by

$$\frac{1}{R^*} = \frac{1}{R_p} + \frac{1}{R_w} \tag{5}$$

where R_p and R_w are the asperity tip radius of the pad and of the wafer, respectively. However, since the wafer surface is assumed to be smooth, R_w is very large compared to R_p, hence $R^* \approx R_p$. Also in Equations (3) and (4), E^* is the equivalent Young's modulus given by

$$\frac{1}{E^*} = \frac{1-\nu_w^2}{E_w} + \frac{1-\nu_p^2}{E_p} \tag{6}$$

where E_p and E_w are the Young's modulus of the pad and of the wafer, respectively; and ν_p and ν_w are the Poisson's ratio of the pad and of the wafer, respectively. Typically E_w is much larger than E_p, so $E^* \approx E_p / (1 - \nu_p^2)$.

From Greenwood's model, the load-balancing equation relating the nominal pressure $\overline{P}$ applied on the wafer surface to the separation distance d [6] is given by

$$\overline{P} = \frac{L_{total}(d)}{A_o} = \frac{4\eta_s E^* \left(R^*\right)^{\frac{1}{2}}}{3} \int_{Z_p>d} (Z_p - d)^{\frac{3}{2}} \varphi(Z_p, t) dZ_p \tag{7}$$

Then the ratio A_f of the real contact area A_{re} to the nominal area A_o is

$$A_f = \frac{A_{re}(d)}{A_o} = \pi\eta_s R^* \int_d^\infty (Z_p - d)\varphi(Z_p) dZ_p \tag{8}$$

By Archard's law [1], the average MRR is proportional to the average real contact pressure P_{re}:

$$P_{re} = \frac{L_{total}(d)}{A_{re}(d)} = \frac{\overline{P} A_o}{A_{re}(d)} = \frac{\overline{P}}{A_f} \tag{9a}$$

$$\overline{MRR} = C_w P_{re} \tag{9b}$$

where the coefficient of proportionality C_w is itself proportional to the relative velocity between the pad and the wafer, and depends on the material properties. Equation (9b) is an alternate form of Archard's law and is similar to Preston's equation [37].

The Greenwood contact model described above assumes that the wafer surface is smooth. However, in most situations, the wafer has a patterned surface with asperities of different heights Z_w, relative to its mean plane. The pad is full of asperities having the properties described earlier in this chapter. Then it follows that the load-balancing Equation (7) can be modified by considering a rough pad surface and an equivalent rough wafer surface [19, 6] with

$$\overline{P} = \frac{4\eta_s E^* \left(R^*\right)^{\frac{1}{2}}}{3} \int_{Z>d} (Z-d)^{\frac{3}{2}} \varphi(Z) dZ \tag{10}$$

where $Z = Z_p + Z_w$ has a PDF $\varphi(Z)$. If Z_w is assumed to have a probability distribution $\psi(Z_w)$, then $\varphi(Z)$ can be estimated. From Equation (10), the load-balancing separation distance d can be solved, provided that $\overline{P}$, $\varphi(Z)$, η_s, E^*, and R^* are given or estimated. In an actual CMP process, the wafer surface height variation is much smaller than the pad asperity roughness; thus, the effect of wafer surface roughness on d can be neglected. Then Equation (10) simplifies to Equation (7).

After the load-balancing separation distance d is known, the mean pressure $\tilde{P}(x)$ at any point on the wafer surface during a small time interval can be calculated as follows. During the time step Δt, many pad asperities make physical contact with the wafer at point x_i, one after the other, assuming that d is constant and that Δt is sufficiently large for the asperities to conform to the height distribution $\varphi(Z_p)$; in other words, if the contact time interval Δt is too small, then the deformable asperities might not have enough time to return to their equilibrium heights, which would result in a different height distribution. If the effect of surrounding wafer topography is ignored, then the mean contact pressure $\tilde{P}$ at point x_i during Δt is equal to the real contact pressure between a rough pad and a locally smooth wafer, which is obtained by using the separation distance $\tilde{d} = d - Z_w(x_i)$ in Equation (9a) [6]:

$$P(x_i) = \frac{L_{total}(d)}{A_{re}(d)} \tag{11}$$

2.2.2. Finite Element Method (FEM) Model of Polishing Pad

The formulation in Equation (9a) and Equation (9b) states that the MRR depends on the average applied pressure and the contact area ratio. However, those two quantities change as the pad deforms mechanically.

Thus, how the pad responds to applied loading via material deformation is important when modeling MRR. The FEM is a well-known, powerful, and popular way to study the physical behavior of a system by computational means, especially when an analytical solution is not attainable.

One of the primary goals of a FEM model and simulation is to obtain an empirical description of a system's behavior, subject to the particular physical laws and constraints that are included in the model, by numerically solving the boundary-value and initial-value partial differential equations. The FEM approach offers the advantage of quantitative analyses and statistical tools. Every FEM model requires some form of external validation.

Relying on the pad and contact models already discussed, the detailed mechanical interactions between the polishing pad, the abrasive particles, and the workpiece surface are modeled and simulated in the ABAQUS/Explicit finite element software, version 6.1 [3], and the validation of the FEM model appears later in this chapter.

Two sets of axisymmetric, finite deformation models simulate (i) the macroscopic response of the long-range asperity contacts for the porous pad, without accounting for the details of the smaller length scale interactions between abrasive particles, and (ii) the local interactions between abrasive particles and the pad. In ABAQUS, the four-node element type (CAX4R) is used.

Three distinct pairs of surface contact interaction are defined: particle-workpiece, particle-pad, and workpiece-pad. The friction between each pair is ignored. The pad is modeled as a porous, hyperelastic material whose uniaxial response is the experimentally measured macroscopic stress-strain graph shown in Figure 7a. The pad's mass density (distinct from the pad's surface density β) is $\rho \approx 544.6$ kg/m^3 and Poisson's ratio is $\nu = 0.15$ [2, 18]. It is important to realize that the polyurethane porous pad has a Poisson's ratio that is much smaller than that of the dense state of polyurethane ($\nu \approx 0.5$). In terms of mechanical response, the workpiece is considered to be rigid in comparison to the soft, porous pad.

2.2.2.1. Long-Range Asperity Contact (Micro-Scale)

The pad asperity model for FEM simulation appears in Figure 10. As discussed earlier in this chapter, the workpiece is represented in the FEM model as a rigid plate. The focus of the model is the macroscopic response of a long-range asperity contact for the porous pad, without accounting for all of the details of interactions between abrasive particles at a finer length scale [3]. In the FEM model of the pad material supporting the asperity, the pad thickness t is the height, and the long wavelength λ of the pad asperity is the diameter. The pad's basic unit of surface roughness, the asperity, is represented by a central, inverted dome region of base length λ_1 and height a. The role played by the asperity is investigated by systematically changing the normalized dome dimensions λ_1 / λ approximately from 0.1 to 1 and a / D_p approximately from 0.1 to 0.5.

The mechanical load in the model is imposed as a displacement boundary condition at the top of the pad throughout the time evolution of the asperity contact, until full contact is made over the entire interval λ [3]. Results from varying the asperity's relative width λ_1 / λ are presented in Figure 11a. It is evident that as λ_1 / λ decreases, the "soft" region of the pad's macroscopic response (displayed in Figure 7a) extends over a longer range of deflection. The similarity of the results for $\lambda_1 / \lambda \approx 0.125$ with those in Figure 7b indicates that $\lambda_1 \approx 100$ μm, which is consistent with the dimension of the central asperity that is marked in Figure 5. Hence, the ratio λ_1 / λ appears to influence the pad's initial contact stiffness. If λ_1 / λ evolves during the CMP process, the MRR will change accordingly, as discussed later in this chapter. Figure 12a shows the results from varying the pad's relative roughness amplitude a / D_p. As a / D_p increases, the "soft" region of the pad's macroscopic response (displayed in Figure 7a) applies over a larger range of deflection. Thus, a / D_p, like λ_1 / λ, affects the pad's initial contact stiffness.

The effect of long-range pad roughness can be further analyzed through a parametric study of the dependence of the normalized contact radius R / λ on the normalized nominal pressure P_o / E_o [3]. If the net nodal force in the axisymmetric FEM model is denoted by F_n, then the applied nominal pressure P_o on a square pad region with size dimension λ is defined by $P_o \equiv F_n / \lambda^2$. Moreover, the FEM results can be approximated by a power-law relation that represents the normalized R versus P_o dependence in the form

$$\frac{R}{\lambda} = \xi_1 \left(\frac{P_o}{E_o} \right)^n, \qquad R < \lambda_1 \tag{12}$$

where ξ_1 is a proportionality constant and n is the exponential order of the power law. Figure 12b shows that the asperity height ratio a / D_p has no appreciable effect on ξ_1 nor n, except for influencing the domain of validity of Equation (12). However, Figure 11b reveals that the asperity width ratio λ_1 / λ modestly affects the power law, since n increases from 0.3 to 0.37 and ξ_1 decreases from 1.35 to 0.71 as λ_1 / λ decreases from 1.0 to 0.1. Equation (12) is an approximate formulation of the asperity contact before full contact between the pad and the workpiece occurs, i.e. the domain of validity is $R < \lambda_1$. For a complete representation that includes full-contact case when $R = \lambda$, an exponential form should be used instead of the power-law form of Equation (12). The FEM simulation uses the R versus P_o dependence given in Equation (12) as the analytical model that characterizes the effect of pad deformation on MRR.

2.2.2.2. Local Abrasive Particle Contact (Nano-Scale)

The FEM model of the localized interaction between an abrasive particle, the pad, and the workpiece appears in Figure 13a; the workpiece is at the top and the pad is at the bottom of the figure [3]. The model shows how the downward pressure exerted by the CMP apparatus is partitioned into the forces transmitted by the abrasive particle to the workpiece and by the pad to the workpiece through direct contact.

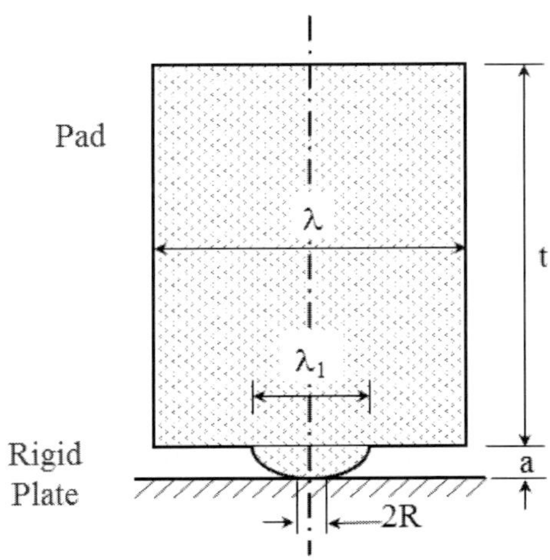

Figure 10. FEM model for long-range asperity contact in porous pad with roughness of large wave-length λ, asperity of smaller wavelength λ_1 and height a, and contact zone of radius R [3].

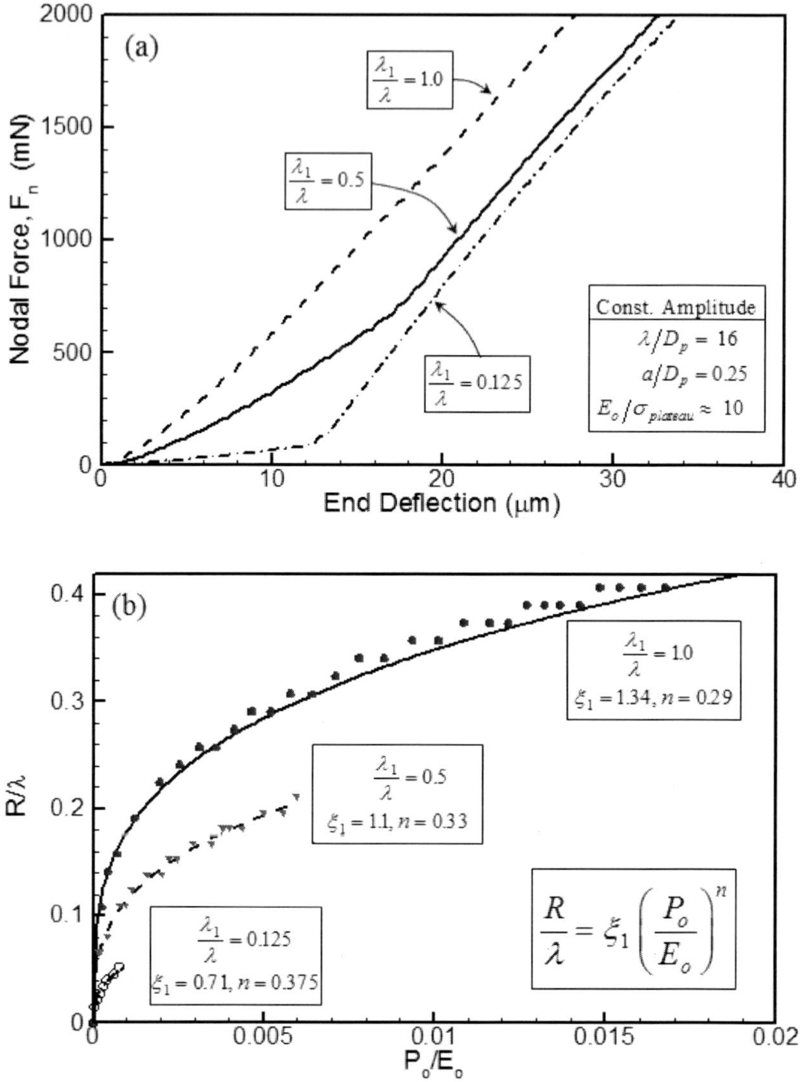

Figure 11. Role of pad asperity morphology ratio λ_1 / λ on initial macroscopic pad response; (a) FEM results of nodal force vs. end deflection; (b) Variation of contact radius R with respect to applied nominal pressure P_o, where data markers are FEM simulation and continuous curve is plot of Equation (12) [3].

The region of interaction between the particle and the pad is set to the inter-particle separation distance L, as labeled in Figure 13b. Several cases are examined via FEM simulation, wherein the particle diameter d is varied from

50 nm to 200 nm, and the corresponding model length L from 500 nm to 2000 nm. Two sets of parametric studies reveal the effect of the abrasive particle size and the effect of the slurry concentration.

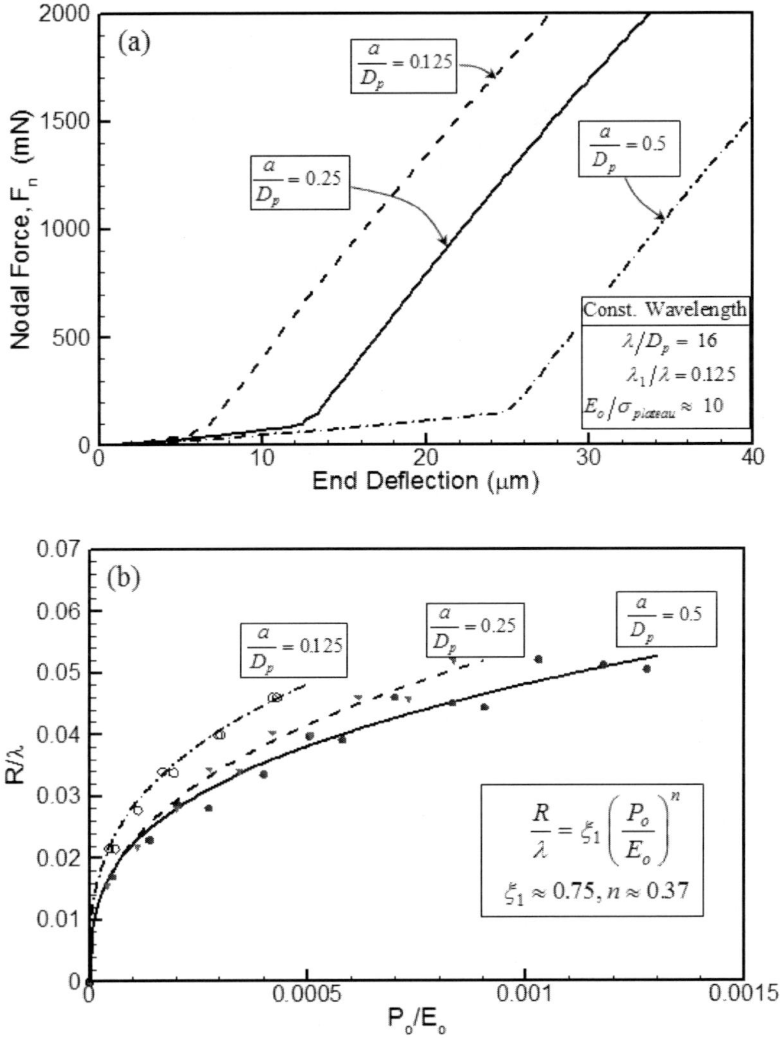

Figure 12. Role of pad asperity amplitude ratio a / D_p on initial macroscopic pad response; (a) FEM results of nodal force vs. end deflection; (b) Variation of contact radius R with respect to applied nominal pressure P_o, where data markers are FEM simulation and continuous curve is plot of Equation (12) [3].

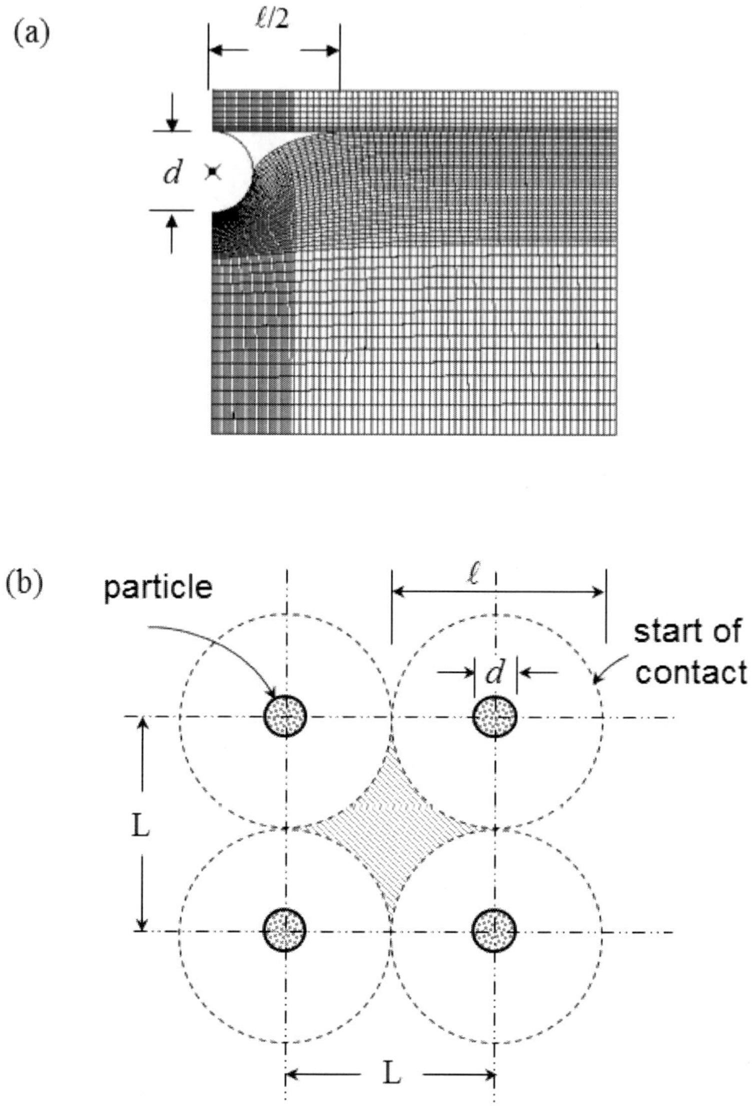

Figure 13. FEM model of pad in vicinity of abrasive particle of diameter *d* and inter-particle spacing *L*; (a) After particle has contacted workpiece; (b) Inter-particle spacing around particles at the onset of pad-workpiece contact [3].

Changing the particle's diameter while keeping the model's length constant, or changing the model's length while keeping the particle's diameter constant, are tantamount to changing the slurry's concentration. For the stated ranges of the parameters' values, the equivalent range for the slurry concentration is from $\alpha \approx 0.67\%$ to $\alpha \approx 2.67\%$.

All cases in the FEM simulation include a model of the pad that is 400 nm thick, which is sufficiently large to isolate the local contact zone around the abrasive particle from the remote boundary effects. The load in the FEM model comes from a boundary displacement condition that is imposed at the top of the workpiece in Figure 13a until there is contact between the workpiece and the particle, which deforms the FEM mesh as shown.

Based on Figure 14a, the effectiveness of the pad in converting the applied local pressure P (distinct from the applied nominal pressure P_o) into a localized force F_p on the abrasive particle is estimated as being related to the characteristic length l, as labeled in Figure 13b, of the pad (immediately surrounding the particle and extending to the nearest neighboring particle) that is not in contact with the workpiece; l may be approximately associated with the diameter of the deformed region of the pad surrounding the particle that does not contact the workpiece [3]. The characteristic length l has an upper bound of L as shown in Figure 14b, which corresponds to no contact between the pad and the workpiece. This case is anticipated for some combination of these conditions: locally stiff pad, short-wavelength asperity, large particle, high particle concentration, or low pressure. The lower bound of l is the particle diameter d, when the deformed asperity envelopes the particle as illustrated in Figure 14d. This scenario may occur for some combination of these conditions: locally soft pad, long-wavelength asperity, small particle, low particle concentration, or high pressure. The value of l may be between d and L for intermediate states, as depicted in Figure 14c.

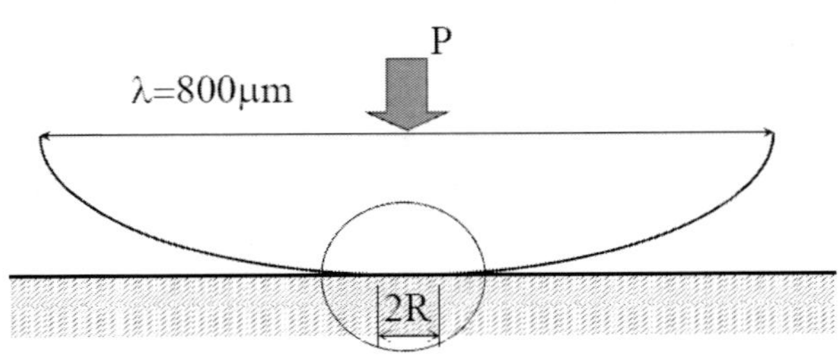

a

Figure 14 (Continued)

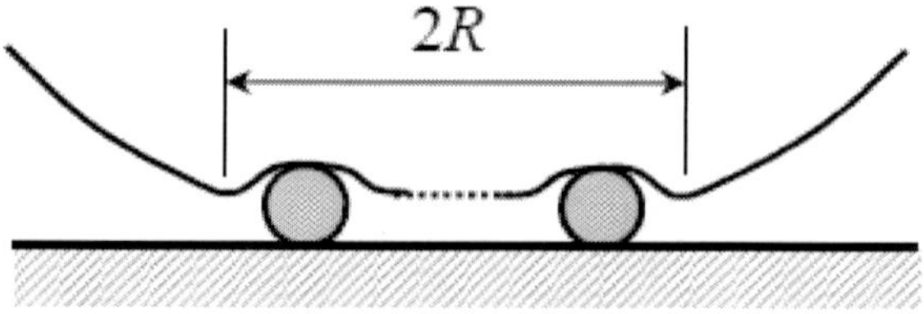

b

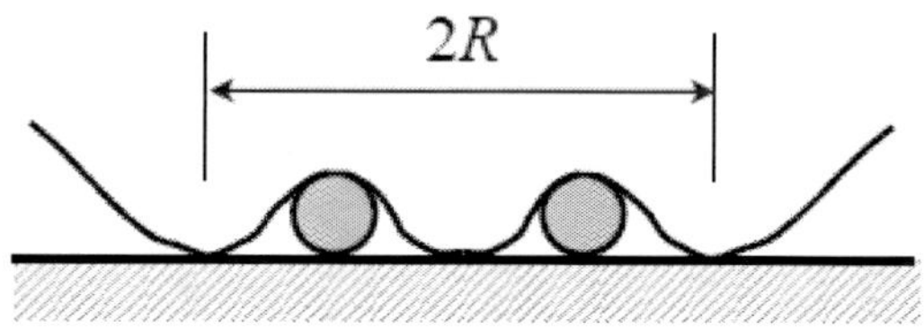

c

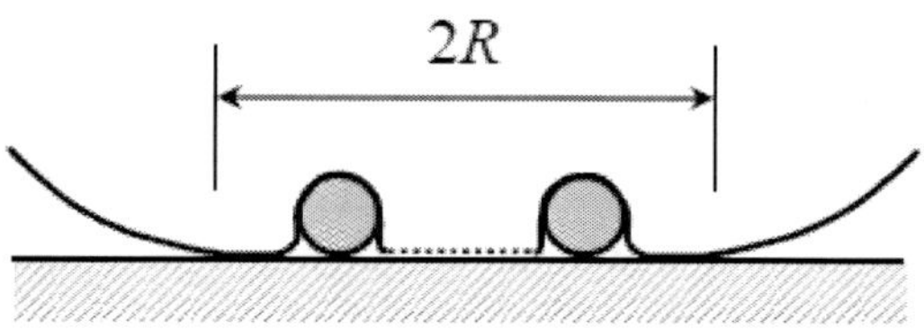

d

Figure 14. Multi-scale contact model with pad on top and workpiece on bottom; (a) Pad-workpiece contact region at macro-scale; (b) No contact between pad and workpiece at nano-scale; (c) Partial contact between pad and workpiece at nano-scale; (d) Full contact between pad and workpiece at nano-scale, adapted from [3].

The dependence of characteristic length l on applied pressure P (distinct from the applied nominal pressure P_o) is plotted in Figure 15a for different values of particle diameter d by holding the slurry concentration α constant and varying the model length L simultaneously with d [3]. On the plot, the markers represent the numerical results, and the continuous curve is the best-fit power-law of the data. The symmetry of the profiles suggests that l scales with d. When l is normalized by d, all of the curves collapse almost to a single curve, as graphed in Figure 15b. A power-law relation can approximately represent such a curve:

$$\frac{\ell}{d} = \xi_2 \left(\frac{P}{E_o} \right)^{-m}, \qquad d < \ell \leq L$$

$$\equiv \left(\xi_2\, \xi_1^{2m}\, \pi^m \right) \beta^m \left(\frac{E_o}{P_o} \right)^{m(1-2n)} \tag{13}$$

where ξ_2 is a proportionality constant, m is the exponential order of the power law, and E_o is the initial elastic modulus of the pad.

Recall that ξ_1 is the proportionality constant and n is the exponential order of the power law in Equation (12), and that β is the average percentage of dense material in an arbitrary cross-section of the pad.

Observe that both ξ_2 and m are nearly independent of both the particle size and the slurry concentration, i.e. $\xi_2 \approx 1.11$ and $m \approx 0.42$. The critical value $(\overline{P})_{NC}$ of the local pressure at which the pad-workpiece contact begins (i.e. when point contact occurs between the pad and the workpiece) is found from Equation (13) by setting $l = L$. Hence $(\overline{P})_{NC}$ is given by

$$(\overline{P})_{NC} = E_o \left(\frac{L}{\xi_2 d} \right)^m \tag{14}$$

The area characterized by l can yield an approximate value of the force F_p that is transmitted to each "active" abrasive particle that is embedded into the pad material, i.e. kinematically trapped by at least one pad asperity [3]. However, due to the edge effect around the particle when pad-workpiece contact begins, the actual value of F_p is slightly higher. To estimate the edge effect at the particle scale, the simulated flat workpiece is pressed against the pad with no entrapped particle.

A plot of the resulting reaction force F_n against the radial coordinate is displayed in Figure 16. The spiked curve is the particle force. The steadily increasing, dotted straight line is the force at the pad-workpiece interface when the particle is omitted.

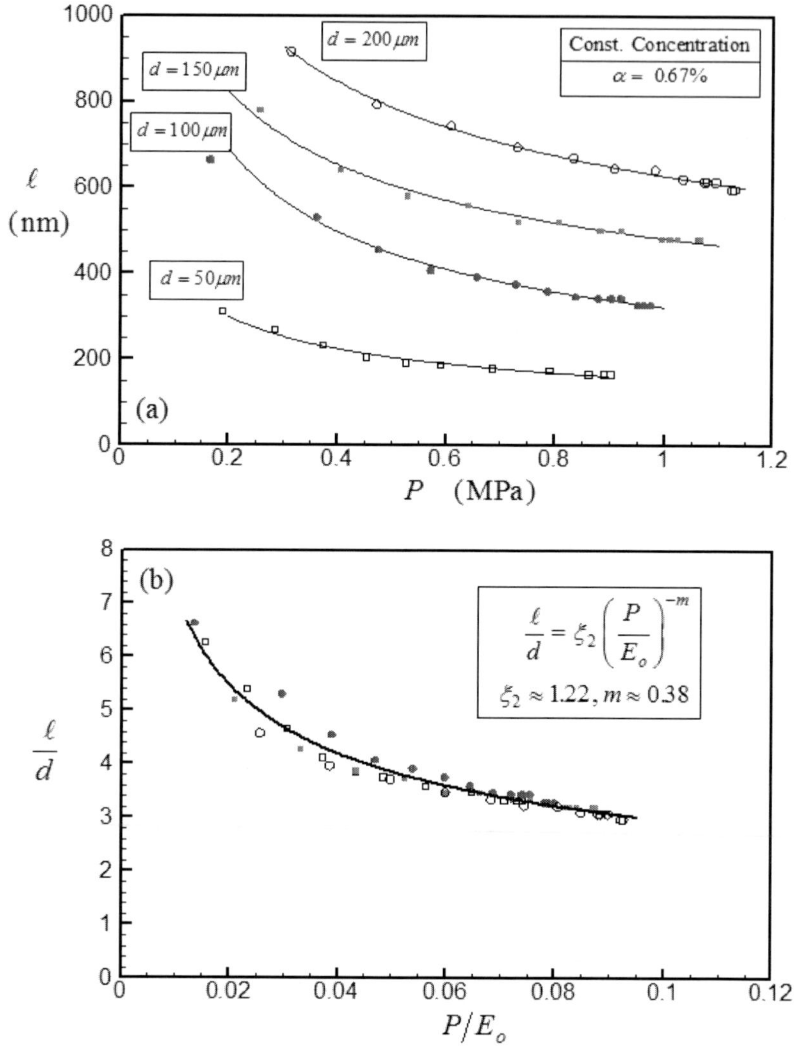

Figure 15. Variation of no-contact region of diameter *l* surrounding each abrasive particle with volumetric concentration α of particles in the slurry kept constant and with appropriate adjustments to model length *L*; (a) *l-P* characteristic curves for different values of particle diameter *d*; (b) Normalized *l-P* characteristic curve approximately showing a power-law dependence [3].

The steadily increasing, connected curve is the force at the pad-workpiece interface when the particle is included. The steady increase in the reaction force with the radial coordinate arises from the increase in the radial area in the axisymmetric model [3].

From this plot, the force transmitted to the particle after pad-workpiece contact has begun can be approximated by

$$F_p = \frac{\pi}{4}\xi_3 \ell^2 \ P, \qquad d < \ell \le L \tag{15}$$

where $\xi_3 \approx 1.55$ accounts for the edge-effect adjustment to the pressure distribution [3].

2.2.3. Physically Based Analytical Model of MRR

The pad asperity contact ratio γ is an important variable that reflects the amplification of the applied nominal pressure inside the contact area due to the presence of asperities which reduce the surface area of pad-workpiece contact [3].

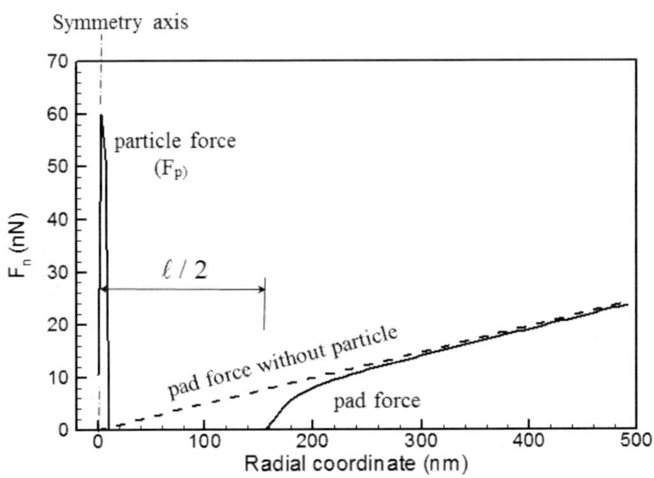

Figure 16. FEM simulation results for local force field, showing how abrasive particles affect the force transmitted from the pad to the workpiece [3].

For a square area of edge length λ, an asperity contact radius R, and a pad surface density β, the asperity contact ratio is given by

$$\gamma = \pi \beta \left(\frac{R}{\lambda} \right)^2 \tag{16}$$

where β is included in Equation (16) to account for the pad's porosity, which is inversely related to the pad surface density [31], within the long-range asperity of wavelength λ. The asperity contact ratio γ defines the local pressure within the pad asperity at equilibrium:

$$P = \frac{P_o}{\gamma} \equiv \frac{P_o}{\pi \beta} \left(\frac{\lambda}{R} \right)^2 \tag{17}$$

For a typical pad morphology, the asperity local pressure is $P \approx O(100P_o)$, i.e. the actual pressure on an asperity tends to be on the order of 100 times the applied nominal pressure due in part to the reduction in contact surface area from the pad's porous microstructure. Figure 17, which is Figure 14 repeated for convenience, shows a pad asperity and three hypothesized contact modes for supporting the applied nominal pressure at the asperity-workpiece interface: (i) no asperity-workpiece contact (Figure 17b), where the applied nominal pressure is transmitted to the workpiece only through the active abrasive particles; (ii) partial asperity-workpiece contact (Figure 17c), where the active abrasive particles transmit a higher percentage of the applied nominal pressure than the asperity due to the details of the contact characteristics around each particle; (iii) full asperity-workpiece contact (Figure 17d), where the applied nominal pressure is supported partially by the asperity and partially by the area fraction occupied by the active abrasive particles [3, 17].

As delineated earlier in this chapter, the applicability of each mode depends on the following factors: the pad's elastic response and surface morphology, the active abrasive particles' size, the slurry's particle concentration, and the applied nominal pressure. The role of each factor is articulated through an analytical expression for the MRR in each of the three contact modes.

In this chapter, the overall model of MRR pertaining to all three contact modes assumes that the slurry has a homogeneous volumetric concentration α

of abrasive particles having a mean diameter *d* and a velocity *u* relative to the workpiece [3].

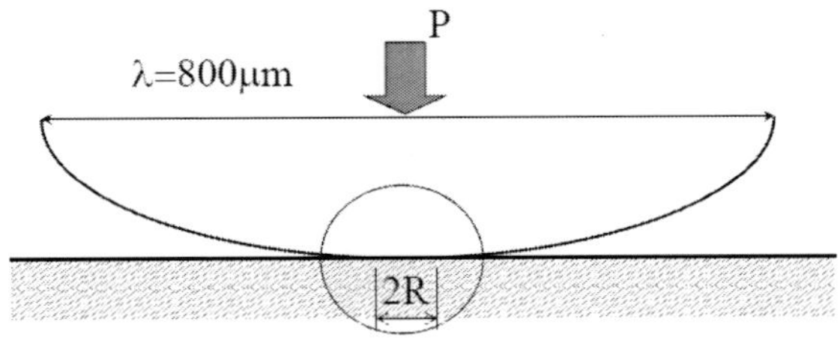

a

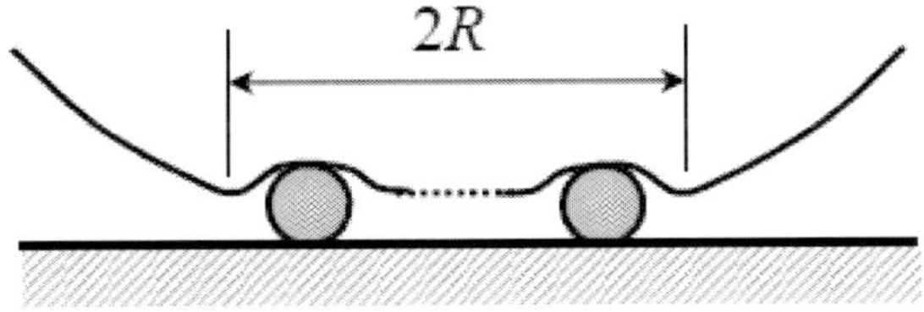

b

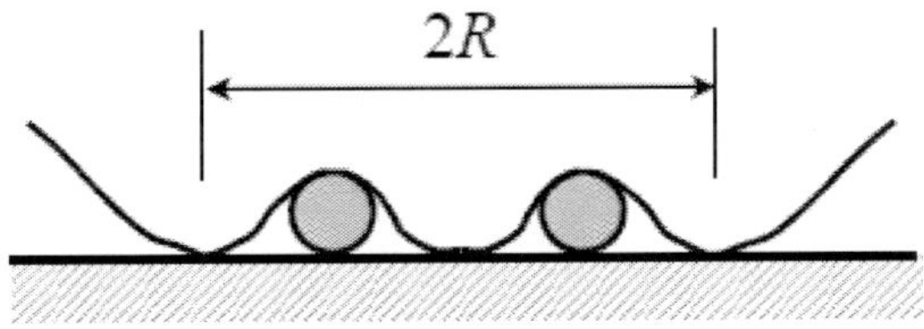

c

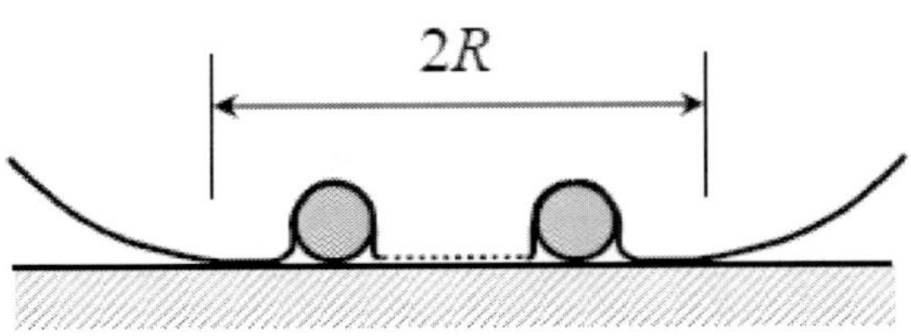

d

Figure 17. (Figure 14 repeated for convenience). Multi-scale contact model with pad on top and workpiece on bottom; (a) Pad-workpiece contact region at macro-scale; (b) No contact between pad and workpiece at nano-scale; (c) Partial contact between pad and workpiece at nano-scale; (d) Full contact between pad and workpiece at nano-scale, adapted from [3].

For a slurry volume of height d, α can be expressed in terms of N, the number of particles per unit area:

$$\alpha = \frac{N\left(\pi\, d^3/6\right)}{d} \equiv \frac{N\,\pi\, d^2}{6} \tag{18}$$

By accounting for the pad surface density (β) and considering only the areas of asperity-workpiece contact (γ), the number of active abrasive particles per unit area is

$$N_a = 6\frac{\alpha\,\gamma}{\pi\, d^2} \equiv 6\,\alpha\,\beta\left(\frac{R}{\lambda\, d}\right)^2 \tag{19}$$

At the particle scale, α can also be expressed in terms of the inter-particle spacing L. From Figure 13b,

$$\alpha = \frac{\pi\, d^2}{6\, L^2} \tag{20}$$

so the inter-particle spacing L can then be expressed in terms of α and d as

$$L = d\sqrt{\frac{\pi}{6\,\alpha}} \tag{21}$$

A phenomenological definition of MRR can be derived by assuming that during a unit time interval, each active particle generates a trench of cross-section area A_c and length u [3, 9]. Then the MRR, which is also the reduction in height of the workpiece surface, caused by the active abrasive particles is

$$MRR = \frac{dH}{dt} \cong N_a\; A_c\; u \tag{22}$$

which explicitly assumes that all of the workpiece material plowed from the surface trenches has been completely detached from the workpiece. This assumption is investigated further in other sources [10].

Because of the small loads that are typical in CMP, the abrasive particles can be assumed to induce small, inelastic surface indentations on the workpiece. Thus, kinematic hardening of the workpiece surface can be ignored, and the entire region of the workpiece surface beneath the abrasive particles can be assumed to be in the yielding state. With these assumptions, the cross-sectional area A_c of the trenches can be calculated by using equilibrium of forces [17], the applied force per particle F_p, and the yield strength of the workpiece material σ_y:

$$A_C \approx \frac{2}{d}\left(\frac{F_p}{\sigma_y}\right)^{1.5} \tag{23}$$

Because the complicated interactions between the chemical phenomena and the mechanical phenomena in CMP are not fully understood [8, 17], it is assumed that the chemical effect is incorporated into the reduction of the yield strength of the workpiece surface (e.g. for oxide CMP), or into the dissolution of the workpiece material plowed from the surface trenches (e.g. for copper or aluminum CMP).

The development of the complete MRR model requires defining the relationship between the applied nominal pressure P_o and the applied force per active abrasive particle in a pad-workpiece contact model [3].

Such a relationship depends on the pad's mechanical properties and surface morphology, and the slurry's properties. Continuing the discussion from earlier in this chapter, the three contact modes between the pad asperities and the workpiece yield three formulations of MRR that are elaborated next, followed by validation of the integrated MRR model comprising the three cases.

2.2.4. Pad-Workpiece Contact Model

Supplementing the overall model of pad-workpiece contact are the particular models of MRR for each of the three contact modes between the pad asperities and the workpiece. The model relationship between the applied nominal pressure and the applied force per active abrasive particle depends on the pad's mechanical properties (e.g. elastic modulus, yield strength) and surface morphology (e.g. pad density, asperity wavelength, asperity-workpiece contact zone radius), and the slurry's properties (e.g. particle concentration, particle mean diameter), among other factors.

2.2.4.1. No Contact between Local Pad Asperity and Workpiece

This case is illustrated in Figure 17b. From equilibrium of forces and Equation (19), the applied force per particle F_p is

$$F_p = \frac{P}{N_a} \equiv \frac{P_o}{6\,\alpha\,\beta}\left(\frac{\lambda\, d}{R}\right)^2 \tag{24}$$

Then, from the application of Equations (12), (19), (22), and (23), the MRR for the case of no contact between the local pad asperities and the workpiece at the interface is

$$\begin{aligned} MRR_{NC} &\cong \left(\frac{2}{\xi_1\sqrt{6}}\right)\frac{u\, E_o^n\, P_o^{1.5-n}}{\sigma_y^{\ 1.5}\sqrt{\alpha\,\beta}}, \qquad P_o < \left(\bar{P}_o\right)_{NC} \\ &\cong 2\frac{u\, E_o^{0.375}\, P_o^{1.125}}{\sigma_y^{\ 1.5}\sqrt{\alpha}} \qquad \text{(for IC1000)} \end{aligned} \tag{25}$$

Equations (14), (16), and (17) permit an estimate of the domain of validity for Equation (25) to be expressed using the critical value of the applied nominal pressure in the non-contact case:

$$\begin{aligned} \left(\bar{P}_o\right)_{NC} &= \gamma\left(\bar{P}\right)_{NC} \\ &\equiv \left(\pi\left(\frac{6}{\pi}\right)^{1/2m} \xi_1^{\,2}\xi_2^{\,1/m}\right)^{\frac{1}{1-2n}} \beta^{\frac{1}{1-2n}}\, \alpha^{\frac{1}{2m(1-2n)}}\, E_o \\ &\cong 20.44\,\alpha^5\, E_o \qquad \text{(for IC1000)} \end{aligned} \tag{26}$$

This means that if the applied nominal pressure is less than the critical value specified in Equation (26), the non-contact model applies and so does the MRR given in Equation (25) [3]. In the absence of asperity-workpiece contact, the MRR expressed in Equation (25) decreases as the slurry's concentration of abrasive particles increases, because if more particles share the applied force that is transmitted through the pad asperities, then each particle experiences less pressure and hence makes a shallower indentation into the workpiece surface, thereby reducing the MRR. However, the MRR in Equation (25) is independent of the particle diameter in this case because the contact model focuses only on the asperity-workpiece contact and not on the particle-workpiece contact. Experimental results and theoretical predictions of the dependence (decrease) of MRR on mean particle diameter are available in other sources [45, 28, 25, 31, 33].

2.2.4.2. Partial Contact between Local Pad Asperity and Workpiece

This case is illustrated in Figure 17c. The applied force per particle F_p is computed by substituting Equations (13) and (17) into Equation (15):

$$F_p = \frac{\pi}{4}\xi_3 \ell^2 P \equiv \left(\frac{\pi^{2m} \xi_2^{\,2} \xi_3}{4 \xi_1^{\,2(1-2m)}} \right) \frac{d^2 E_o^{\,2(n-2nm+m)} P_o^{\,1-2(n-2nm+m)}}{\beta^{(1-2m)}}, \quad d < \ell \le L \tag{27}$$

Then the MRR for the case of partial contact between the local pad asperities and the workpiece at the interface is

$$MRR_{PC} \cong \left(\frac{3\pi^{3m} \xi_2^{\,3} \xi_3^{1.5}}{2 \xi_1^{\,(1-6m)}} \right) \bullet \frac{u\, \alpha\, E_o^{(n-6nm+3m)} P_o^{\,1.5-(n-6nm+3m)}}{\sigma_y^{\,1.5}\, \beta^{(0.5-3m)}}, \qquad P_o \ge (\bar{P}_o)_{NC}$$
$$\cong 6.4 \frac{u\, \alpha\, E_o^{0.66} P_o^{0.84}}{\sigma_y^{\,1.5}} \qquad \text{(for IC1000)} \tag{28}$$

When there is partial asperity-workpiece contact, the MRR modeled by Equation (28) has a linear dependence on the slurry's particle concentration, which is very different from the MRR modeled by Equation (25) when there is no asperity-workpiece contact [3]. Another significant difference between the non-contact and the partial-contact cases is the exponential order of the power-law relationship between MRR and applied nominal pressure P_o. Moreover, the pad surface density β now affects MRR. The partial-contact MRR is independent of the abrasive particle diameter for the same reasons and with the same caveats as discussed in the non-contact scenario earlier in this chapter.

2.2.4.3. Full Contact between Local Pad Asperity and Workpiece

This case is illustrated in Figure 17d. From equilibrium of forces, the applied force per particle F_p is the product of the particle's cross-sectional area and the applied local pressure:

$$F_p = \frac{\pi}{4} d^2 \; P \equiv \frac{P_o}{4\,\beta}\left(\frac{\lambda\, d}{R}\right)^2 \tag{29}$$

Then, from the substitution of Equations (29), (12), and (19) into Equation (22), the MRR for the case of full contact between the local pad asperities and the workpiece at the interface is

$$MRR_C \approx \left(\frac{3}{2\,\xi_1}\right)\frac{u\,\alpha\,E_o^n\,P_o^{1.5-n}}{\sigma_y^{1.5}\,\sqrt{\beta}}, \qquad P_o \geq \left(\bar{P}_o\right)_C \tag{30}$$

It is interesting that the exponential order of the power-law dependence of the MRR on the applied nominal pressure is the same in the full-contact case as in the non-contact case [3].

Therefore, if the change from non-contact mode to full-contact mode occurs while MRR stays constant, the ratio of the critical values of the applied nominal pressure between the two cases is

$$\frac{\left(\bar{P}_o\right)_C}{\left(\bar{P}_o\right)_{NC}} = \left(\frac{1}{3\alpha\sqrt{6\,\alpha}}\right)^{\frac{1}{1.5-n}} \tag{31}$$

where $\left(\bar{P}_o\right)_{NC}$ is given in Equation (26). For typical values of pad stiffness and slurry particle concentration in CMP, the critical pressure $\left(\bar{P}_o\right)_C$ for attaining the full-contact mode is generally at least two orders of magnitude larger than the critical pressure $\left(\bar{P}_o\right)_{NC}$ for attaining the partial-contact mode.

2.2.5. Validation of MRR Model

The physically based analytical model of the MRR in CMP that has been constructed in this chapter will now be validated by comparing its predictions of MRR against measurements of MRR obtained from experiments under various conditions. Previously in the chapter, the models based on the hypothesized mode of physical contact between the pad asperities and the workpiece specify the variation of the MRR with respect to the applied nominal pressure P_o, the pad's stiffness E_o and morphology parameter β, and the slurry's volumetric concentration α of abrasive particles [3]. If the pad is not reconditioned during CMP and all other factors are held constant, the pad morphology will evolve over time toward less surface roughness, and the MRR will decrease as the CMP process runs [33, 45, 28, 25, 31]. For this reason, the comparison between model and experiment presumes a "fresh" pad and examines the effects of P_o and α. For each of the three modes of asperity-workpiece contact, the model's predicted MRR curve is calibrated to the experimental data using a single data point in order to account for unknown experimental parameters that are not reported in the referenced articles.

The dependence of MRR on applied nominal pressure is verified against empirical data obtained from oxide CMP [34]. The experiment is conducted on the Avante™ 472 polishing machine from IPEC Planar/Westech using the SS25 commercial slurry from Cabot Corporation at a flow rate of 900 ml/min [3]. Two dual-layered pads, with either the IC1000 or the IC1400 as the top pad and the Suba IV as the lower pad, perform the polishing. The dual-layered pads have the same top-layer stiffness and morphology, referring to the discussion from earlier in this chapter. The downward nominal pressure applied by the carrier varies from 4.8 psi (33.1 kPa) to 7.2 psi (49.6 kPa), the platen's rotational speed ranges from 20 rpm to 80 rpm, and the offset between the platen's center and the workpiece's center is 170 mm.

The experimental data points in Figure 18 exhibit two major regimes of behavior [3]. Note that these data points correspond to different relative velocities between the pad and the workpiece; each group of observations at the same relative velocity is marked by a different symbol.

The model's prediction of MRR uses the following parameter values from a typical IC1000 pad: D_p = 50 μm, E_o = 180 MPa, $\lambda / D_p = 16$, $\lambda_1 / \lambda = 0.125$, and $a / D_p = 0.25$. The partial-contact prediction is the bold, solid curve; the non-contact prediction is the dashed line; and the full-contact prediction is the dashed-dotted line. Figure 18 shows that the model's prediction fits well with the experimental observations. The empirical measurements reveal a change in slope that occurs roughly at a pressure somewhere between 1.0 kPa and 4.0 kPa, and at a relative velocity somewhere between 85 m/min and 21 m/min. The model predicts that a transition from no contact to partial contact between the pad and the workpiece occurs approximately at a pressure somewhere between 0.37 kPa and 1.125 kPa, and at a slurry volumetric particle concentration somewhere between 4% and 5%. For a lower concentration, the transition from non-contact to partial contact would occur at a much lower pressure level [3]. From Equation (31) with α = 5%, $\left(\bar{P}_o\right)_C \approx 12\left(\bar{P}_o\right)_{NC}$. These two critical values of the applied nominal pressure, which govern transitions between the three pad-workpiece contact modes described earlier in this chapter, are marked in Figure 18. It is apparent that a transition from the non-contact mode or the partial contact mode to the full contact mode is not feasible if the MRR is held constant; instead, evidently a much higher applied nominal pressure is required for such a transition to occur.

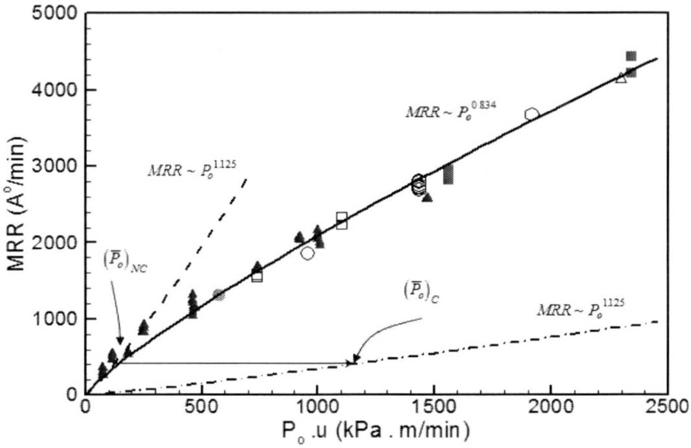

Figure 18. Comparison of model prediction with experimental data for dependence of MRR on applied pressure in oxide CMP [34]; groups having similar relative velocities are marked by same symbol [3].

The dependence of MRR on abrasive particle concentration in the slurry is verified against empirical data obtained from BK7 glass CMP using cerium oxide ("ceria") abrasive particles [3].

Figure 19 shows the model's prediction for MRR (bold, solid curves) along with the experimentally measured data points, which show that the MRR initially jumps sharply from 0 μm/min to 0.3 μm/min as the ceria concentration increases from 0% to 5%, then it plateaus around 0.3 μm/min as the ceria concentration rises from 5% to 30%, and then it steadily declines to about one-third of the maximum polishing rate as the ceria concentration grows from 30% to 70% [23].

In the low and intermediate concentration regimes (roughly from 0% to 30% in Figure 19), the model considers full contact and partial contact, respectively, between the pad asperities and the workpiece, which implies a linear relationship between MRR and α as expressed in Equation (30) and Equation (28), respectively [3].

In the high concentration regime (roughly from 30% to 70% in Figure 19), the model assumes no direct contact between the pad asperities and the workpiece, so the MRR is expected to vary according to $1/\sqrt{\alpha}$ as indicated in Equation (25).

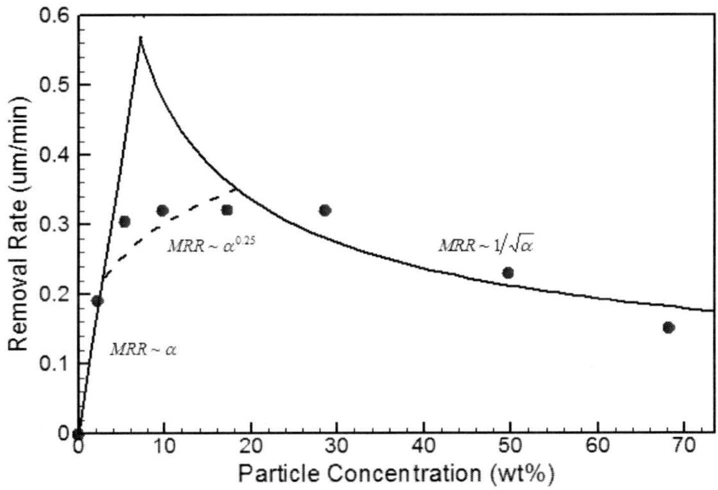

Figure 19. Comparison of model prediction with experimental data for dependence of MRR on particle concentration in BK7 glass polishing [23, 3].

The two types of relationship with respect to concentration are very different; hence, in the transitional regime between low concentration and high concentration, an interpolation between the modeled relationships is expected, which is why the best-fit power-law relationship between MRR and α is graphed as the dashed line to serve as a reference. Figure 19 shows that the model's prediction reasonably fits with the experimental data; the modeling error may be related to the unusual properties of glass arising from its special thermodynamic state.

3. Development of Slurry Particle Agglomeration Model

Abrasive particles suspended in slurry tend to coagulate/agglomerate as governed by the Smoluchowski equations, which give a volumetric description of a physical process using a statistical formulation [40, 22, 25]. The agglomeration process is determined by the pH of the slurry, which drives the surface charge on the slurry particles and hence their ability to agglomerate. The differential equation that defines the rate at which the number N of particles having volume M is changing with respect to time [25, 4, 7] is

$$\frac{d}{dt}N(M) = \frac{1}{2}\sum_{K=1}^{M-1} a(M,K)N(M-K)N(K) - \sum_{K=1}^{\infty} a(M,K)N(M)N(K) \tag{32}$$

where $a(M,K)$ is the agglomeration kernel function that governs the kinetics and speed of this process that is given by

$$a(M,K) = \frac{8k_B T N_0}{3\eta} \frac{(M^{1/3}+K^{1/3})^2}{M^{2/3}+K^{2/3}} e^{-\frac{U(r)}{k_B T}} \tag{33}$$

The first term on the right-hand side of Equation (32) is the rate of increase in the number of particles having volume M that occurs when particles having volumes smaller than M agglomerate to form particles of volume M.

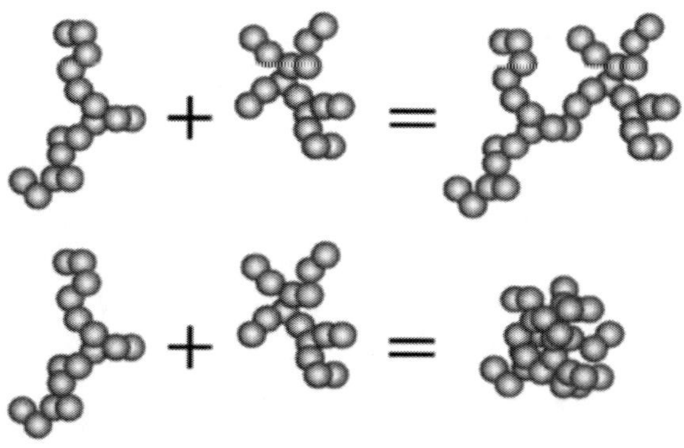

Figure 20. Different possible structures resulting from particle agglomeration, adapted from [25].

The second term on the right-hand side of Equation (32) is the rate of decrease in the number of particles having volume M that occurs when particles having volumes of M participate in agglomeration to form particles having volumes larger than M. An illustration of particle agglomeration appears in Figure 20. In Equation (32) and Equation (33), the volumetric quantities M and K are statistical distributions in the sense that their values are taken from their respective probability distributions, i.e. they are continuous random variables [25]. On the continuous interval of values of M, the frequency function $N(M)$ is a continuous probability distribution (PDF).

The agglomeration kernel $a(M,K)$ in Equation (33) is dependent on the Boltzmann constant k_B, the temperature T of the slurry, the viscosity η of the slurry, the initial number of particles N_o per unit volume, and the electrostatic potential energy $U(r)$ between two repelling particles. When the pH of the slurry is at or around the particles' iso-electric point, the slurry particles carry no surface charge, so there are no electrostatic repulsive forces keeping them apart; when they are close enough to one another, van der Waal forces bind them together [25, 7].

In this situation, the exponential factor in Equation (33) becomes unity, the particle surface area factor involving M and K becomes unity because agglomeration becomes independent of surface area, and N_o becomes N because agglomeration becomes dependent on the actual number of particles per unit volume at any given instant of time, instead of the initial number of particles per unit volume. Therefore, when the pH of the slurry is at or around the iso-electric point of the slurry particles, the result is a large kernel value corresponding to a faster agglomeration speed that is given by

$$a(M,K) = \frac{8k_B TN}{3\eta} \tag{34}$$

This regime of particle agglomeration is termed diffusion limited agglomeration (DLA) since the process of particle agglomeration is limited only by the diffusion process (e.g. Brownian Motion) for the particles [7, 22; 25].

On the other hand, if the pH of the slurry is not near the iso-electric point of the abrasive slurry particles, the particles carry a net surface charge that causes electrostatic repulsion between them, which decreases the likelihood of collisions or near-collisions that are required for agglomeration to occur [7, 25]. The abating exponential term in Equation (33) becomes non-negligible, resulting in a small kernel value corresponding to a slower agglomeration speed. This regime of particle agglomeration is termed reaction limited agglomeration (RLA) since the process of particle agglomeration is limited by the net surface charge on the particles, which in turn is dependent on the slurry pH, which is related to the chemical reactions that occur during CMP. DLA occurs more rapidly than RLA because electrical repulsion limits the process much more than diffusion does.

The effect of pH on slurry particle agglomeration can be seen in numerical simulation results. Figure 21 shows the absence of agglomeration for silica particles whose iso-electric point is 3, in a slurry whose pH is 7 [42]. This is the RLA scenario in which the slurry particles carry a net electrical charge on their surfaces. The repulsive interactions inhibit agglomeration, causing the particles' Gaussian size distribution (mean 240 nm, standard deviation 30 nm) to remain relatively unchanged over time [25]. Figure 22, on the other hand, shows aggressive agglomeration for silica particles in a slurry whose pH is 3. This is the DLA scenario in which the slurry particles do not have an appreciable electrical charge on their surfaces. With no electrically repulsive interactions, the particles are free to diffuse and to agglomerate, as evidenced by the tail of the particle size distribution in Figure 22b, which shows growth in probabilistic particle volume (hence, size) with respect to time. The maximum particle volume is 4 μm^3 in the simulation because the computational resources required for analyzing higher volumes are limited. Figure 22b shows that the probability of existence of larger particles, which can produce deeper indentations in the workpiece surface during CMP, increases over time, resulting in an increase in defectivity; a more detailed explanation appears later in this chapter.

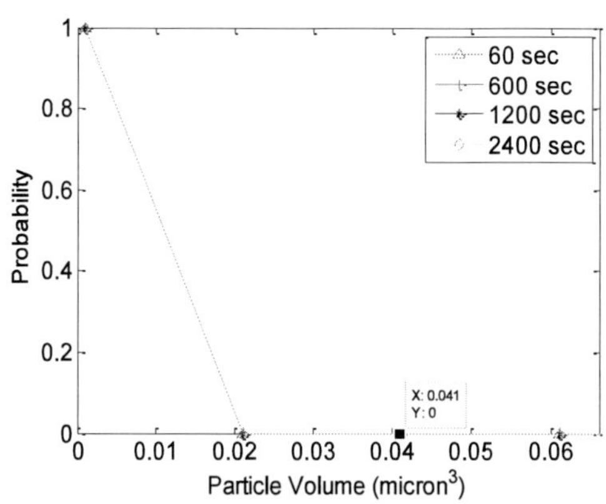

Figure 21. Time evolution of particle size distribution for silica particles in the RLA case, showing virtually no change (superimposed data markers) [25].

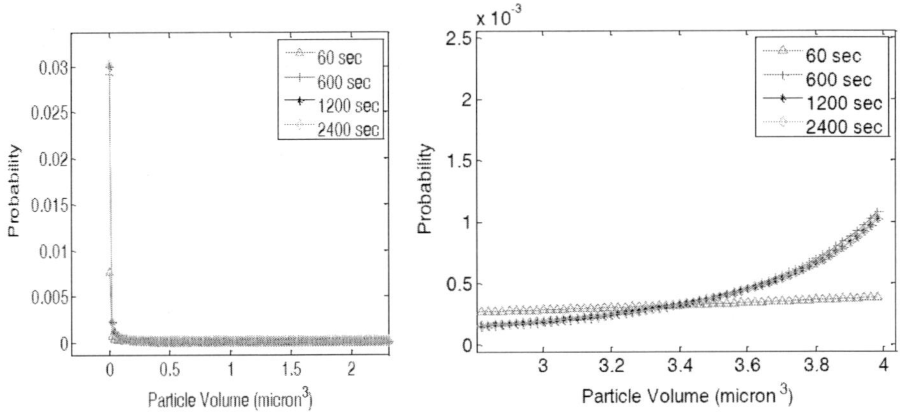

(a) (b)

Figure 22. Time evolution of particle size distribution for silica particles in the DLA case; (a) Process occurs rapidly and equilibrates within 60 seconds; (b) Tail of particle volume distribution shows agglomeration while CMP runs [25].

The speed of agglomeration directly drives the size of the particles, which in turn drives the size of scratches formed on the workpiece. Hence, the pH of the slurry solution has a direct effect on the defectivity of the workpiece.

In CMP, traditional slurries comprising abrasive particles in a chemical suspension are susceptible to agglomeration phenomena (for example, driven by the pH level) that worsen the scratch performance of the process.

Abrasive-free slurries may be a feasible solution to the agglomeration problem that exacerbates scratches on the finished surface. For instance, aqueous solutions of oxalic acid and hydrogen peroxide are viable abrasive-free slurries for CMP of copper [35], in which chemical effects dominate mechanical effects during the removal of copper surface layers, since colloidal silica abrasive particles have no measurable effect on the relatively high polish rates of copper.

The mechanism of copper removal in abrasive-free solutions of H_2O_2 and oxalic acid is affected by pH and H_2O_2 concentration [35]. Four dicarboxylic acids (oxalic, malonic, succinic, and glutaric, in order of increasing carbon chain lengths) are possible complexing agents for a chemically dominated CMP process [38].

Copper dissolution and polishing are correlated with pH-dependent distributions of mono-anionic (oxalic and malonic) and neutral (succinic and glutaric) acid species [38]. CMP processes employing abrasive-free slurries such as these acids produce less defectivity and better planarity when compared to traditional CMP processes employing abrasive slurries [38]. Further information on abrasive-free slurries is available from other sources [49].

4. Development of Defectivity Model for Scratch Propensity

Many different models for material removal in CMP have been proposed. Most, if not all, of the models include scratching of the workpiece surface (or the layer undergoing CMP) as an essential physical phenomenon that is involved in the material removal process [7].

For instance, the scratch intersection model proposed by Che assumes that material is removed only when a scratch generated by a particle intersects with a previously formed scratch [9]. The scratch intersection model proposes that there is a characteristic length that is related to the amount of material removed by the CMP process, and it has been verified by experiments. There are alternative models that assume material is removed whenever a scratch is generated by a particle [26], and this class of models is used in the remainder of this chapter to quantify material removal in CMP [25].

The physical phenomenon of scratch generation is a necessary bane in the CMP process.

On the one hand, the absence of scratches precludes material removal, yet on the other hand, the presence of a scratch that is too deep (compared to an application-specific threshold) may render the entire product (e.g. silicon wafer costing many tens of thousands of dollars) non-functional. Hence, it is valuable to have a model of the scratch formation process that can quantify the scratch distribution, which is the probability distribution (PDF) of scratch depth, in terms of pad parameters, slurry parameters, and process parameters. The scratch propensity is defined as the probability of occurrence of the deepest scratch under consideration [25].

Based on the discussions of pad evolution (asperity erosion) and slurry evolution (particle agglomeration) earlier in this chapter, as the pad wears down during the material removal process, more asperities come into contact with the workpiece.

The increase in surface area of contact reduces the maximum asperity-workpiece contact pressure at the interface, which improves (decreases) scratch propensity, but hurts MRR, which lengthens the time needed for a given CMP process [28, 25, 45, 31, 33].

The scratch probability distribution is affected by two stochastic random variables: the pad asperity height probability distribution f_Z, and particle diameter probability distribution f_D [25]. The indentation depth w of a scratch that is generated by the jth particle under the ith asperity [26, 25, 7, 4] is given by

$$w(i,j) = \frac{2F(i,j)}{\pi H_w d(i,j)} = \frac{P_{cont}(i)}{2H_w} d(i,j) = \frac{2E\sqrt{\kappa_s}}{3H_w \pi^2} d(i,j)\sqrt{Z_i - s(t)} \tag{35}$$

where F is the force load exerted on the jth particle by the ith asperity, H_w is the hardness of the workpiece, d is the particle diameter, P_{cont} is the pressure exerted on the jth particle by the ith asperity, E is the elastic modulus of the pad, κ_s is the curvature of the asperity tip, Z_i is the height of the ith asperity, and $s(t)$ is the mean pad-workpiece separation distance as a function of time.

The particle distribution N in the agglomeration model expressed in Equation (32) is a function of a volumetric distribution, the random variable M.

In order to define the scratch distribution, the particle volume distribution $N(M)$ must be mathematically transformed into the particle diameter distribution $f_D(d)$ [25]. The transformation from a diameter probability distribution $f_D(d)$ into a volume probability distribution $f_V(v)$ is

$$f_V(v) = \frac{1}{3}\left(\frac{6}{\pi v^2}\right)^{\frac{1}{3}} f_D\left(\left[\frac{6v}{\pi}\right]^{\frac{1}{3}}\right) \tag{36}$$

so the inverse transformation from a volume distribution into a diameter distribution is

$$f_D(d) = \frac{\pi d^2}{2} f_V\left(\frac{\pi d^3}{6}\right) \tag{37}$$

Substituting N for f_V in Equation (37) transforms $N(M)$ into $f_D(d)$:

$$f_D(d) = \frac{\pi d^2}{2} N\left(\frac{\pi d^3}{6}\right) \tag{38}$$

Then the cumulative distribution function $F_W(w)$, which gives the probability that a scratch has depth less than or equal to w [25, 7], is

$$F_W(w) = \int_0^{D_{\max}} \int_0^{C} f_D(d) f_{\bar{Z}}(z) d(d) d\bar{z} \tag{39}$$

where $\bar{z} = z - s$, $C = \left(\frac{3wH_w\pi^2}{2E\sqrt{\kappa_s}}\right)^2$, D_{max} is the largest value of particle diameter under consideration, and $d(d)$ denotes the infinitesimal differential of the particle diameter for the purpose of integration. Lastly, the scratch propensity or scratch distribution (PDF) $f_W(w)$, which gives the probability that a scratch has depth equal to w, is obtained by differentiating Equation (39) with respect to scratch depth:

$$f_W(w) = \frac{\partial F_W}{\partial w}(w) \tag{40}$$

Using Equation (40), the scratch distribution is calculated for a workpiece of modulus 70 GPa, a layer to be removed of hardness 4 GPa, a nominal pressure of 34 kPa, and a slurry having parameters as described earlier in this chapter.

The scratch distribution plotted in Figure 23 is for the case of slurry pH = 3, i.e. the DLA regime. It is evident from the plot that scratch formation is a result of aggressive agglomeration of slurry particles in the DLA case. The probability of a deep scratch and the scratch propensity increases with respect to time [25]. The maximum volume of the particles in the simulation is discussed earlier in this chapter, and its effect is that the scratch depth appears to reach a limit of 40 nm in Figure 23. The much faster growth in the average effective particle size due to DLA, compared to that due to RLA, increases scratch propensity and worsens scratch performance.

Now the effect of slurry particle agglomeration on scratch propensity can be examined. As discussed earlier in this chapter, there are different types of

CMP processes for different types of workpieces, layers, and applications. For example, oxide CMP planarizes the workpiece after trenches are made in the workpiece for the subsequent fabrication of conductive pathways (i.e. metal) in circuits. Another example is copper CMP, which planarizes the workpiece after the metal is deposited into the trenches in order to remove the excess metal ("overburden") resulting from overfill, resulting in a patterned layer of conductive "lines" for the subsequent fabrication of the next layer in a multi-layer circuit. In the example of copper damascene process, oxide CMP is a precursor to copper CMP. The local MRR in oxide CMP is affected by the local pattern density on the workpiece surface. The metallic conductor forming the circuit is laid once the pattern is made, and then copper CMP is performed.

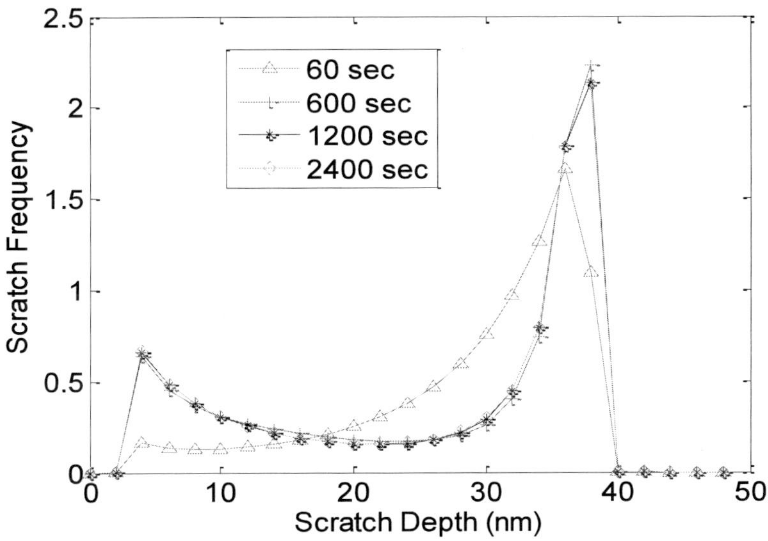

Figure 23. Scratch distribution for DLA case.

The effects of pattern density on MRR may be studied using the "MIT Mask" that was created by Ouma at the Massachusetts Institute of Technology for the purpose of CMP process characterization [34]. The mask occupies an area of 20 mm x 20 mm, divided into 25 square zones of area 4 mm x 4 mm. Within each zone, the pitch (the average distance between corresponding locations on successive lines, i.e. the average wavelength of the pattern profile) and the pattern density (the average ratio of line width to pitch) are constant and uniform. The zones are arranged in a specially designed layout, whose spatially varying pattern density (e.g. "PD 0.5") and pitch (e.g. "P 30 micron") are shown in Figure 24. Using the analytical models in this chapter, the probability of scratch formation in each zone of the MIT Mask can be found via computational simulation. For example, the likelihood of not having a scratch deeper than 4 nm after 600 seconds of CMP for a particular test scenario also appears in Figure 24 [47, 6].

Using the scratch propensity model in this chapter, the probability inside each zone of not having a "catastrophic scratch" (deeper than 4 nm, for example) after 600 seconds of CMP can be computed. It can be seen from the result shown in Figure 24c that the probability of catastrophic scratches is higher, hence the scratch propensity is higher, in zones having a higher pattern density. This is because an active abrasive particle, which is larger than the hydrodynamic layer between the workpiece and the pad, is more likely to generate a scratch as the pattern density increases. These results, taken with the results discussed earlier in this chapter, show that diffusion-limited agglomeration of slurry particles increases scratch propensity and degrades scratch performance much more than reaction-limited agglomeration.

5. Effects of Consumables on Material Removal and Scratch Propensity

After the CMP process has been modeled and tested, it can be better controlled and optimized. One of the primary strengths of computer-based simulation is the ability to conduct virtual experimentation on a physical process, which could save money, time, and human labor, among other benefits. Once modeled, programmed, and experimentally validated on benchmark cases, such a computational simulation platform permits changes to be made relatively easily, variables to be isolated, and simulations to be run in an iterative or in a parallel manner.

The models presented in this chapter for MRR and scratch propensity in CMP have been implemented as a computational simulation platform called "CMPSim" (http://www.cmpsim.com) whose results compare well with experimental results [25, 6, 47].

Figure 25 shows a representative set of predictions from CMPSim that yield small errors, which are summarized in Table 2, relative to experimental measurements [34]. First, the simulation omits the particle agglomeration

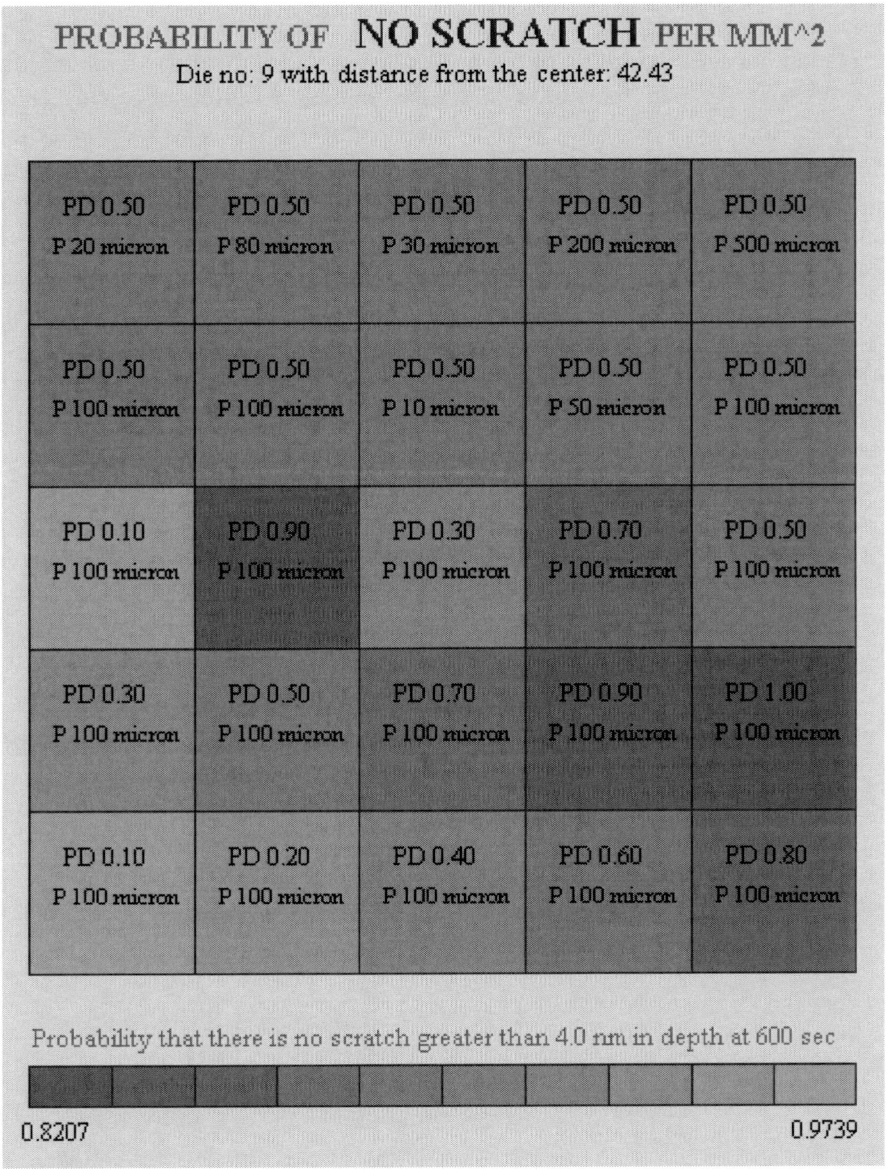

Figure 24. 20 mm x 20 mm MIT mask layout with probability of no scratches per mm^2 inside each 4 mm x 4 mm square zone [25].

model, and then the RLA and DLA models are separately included to predict the surface profile of the wafer. For compatibility with Ouma's empirical data, the software makes predictions after 29 seconds and after 88 seconds of simulation. In Figure 25, the plots in the top row do not account for pad relaxation nor time evolution of the pad or the particles; the plots in the middle row account for pad relaxation and time evolution of the pad with DLA particles; and the plots in the bottom row account for pad relaxation and time evolution of the pad with RLA particles. In Figure 25, the plots in the left column pertain to the "up" area on the wafer, e.g. metal lines, whereas the plots in the right column pertain to the "down" area on the wafer, e.g. insulator trenches.

The simulation results indicate that CMPSim and the associated models addressed in this chapter accurately reflect how pattern density affects the surface profile of the MIT Mask during CMP [47]. Based on Figure 25, where the pattern density is lesser, the height of "up" area is polished at a faster rate, and where the pattern density is greater, the height of the "up" area is polished at a slower rate.

Based on Table 2, the RLA model is better than the model without particle agglomeration, which in turn is better than the DLA model. This inference is reasonable if the slurry used by Ouma was stabilized. Moreover, based on the empirical results after 88 seconds of CMP, it seems that RLA was in effect, rather than DLA.

There are many variables pertaining to consumables used in CMP that could be included in a parametric study on their effects on MRR and scratch propensity. The parametric study in this chapter focuses on a few of the main variables in order to demonstrate how certain choices for the consumables can improve the CMP process in terms of MRR and scratch propensity.

Table 2. Small errors between CMPSim simulation and experimental results using MIT Mask [47]

Line Number on Wafer	RMS Prediction Error (Angstrom), No Agglomeration, No Pad Evolution	RMS Prediction Error (Angstrom), DLA, Pad Evolution, Pad Relaxation	RMS Prediction Error (Angstrom), RLA, Pad Evolution, Pad Relaxation
L3	396	420	347
L4	418	468	379
L5	434	448	373

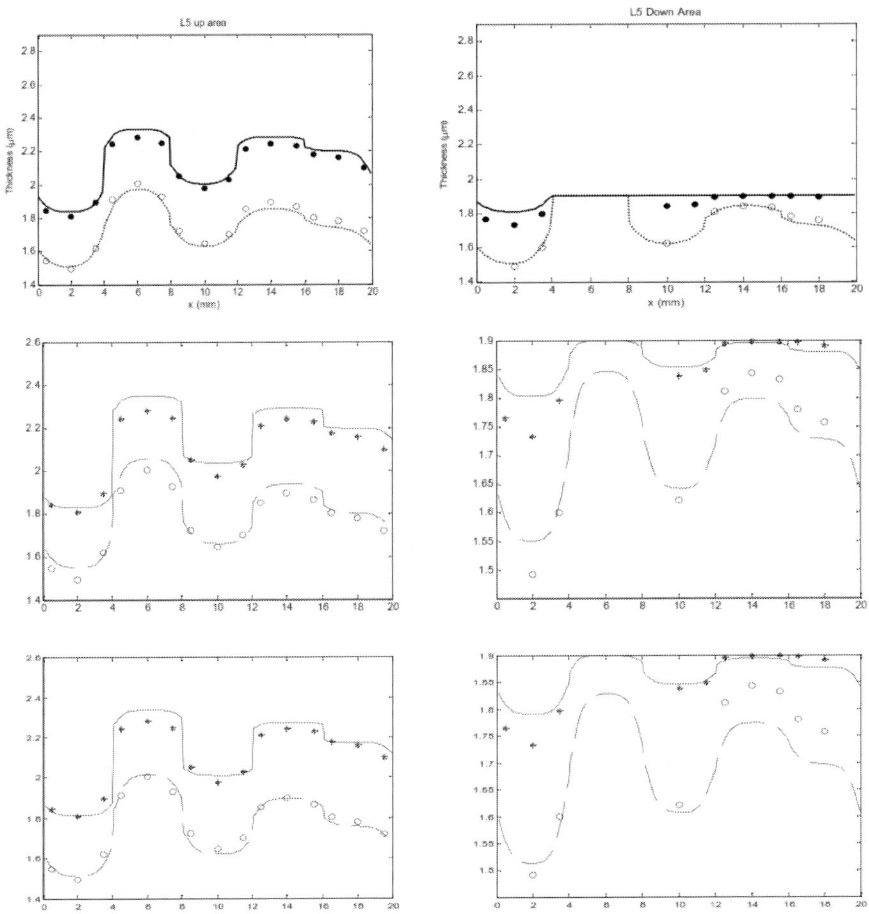

Figure 25. Close agreement between CMPSim simulation and experimental results using MIT Mask; horizontal axis is position (x) in mm, vertical axis is thickness in µm; top row does not model pad relaxation nor evolution of pad or particles; middle row models pad relaxation and evolution of pad and DLA particles; bottom row models pad relaxation and evolution of pad and RLA particles; left column contains plots for "up" area, e.g. metal lines; right column contains plots for "down" area, e.g. insulator trenches; solid lines are CMPSim predictions and filled data markers are experimental observations after 29 seconds of CMP; dashed lines are CMPSim predictions and open data markers are experimental observations after 88 seconds of CMP [47].

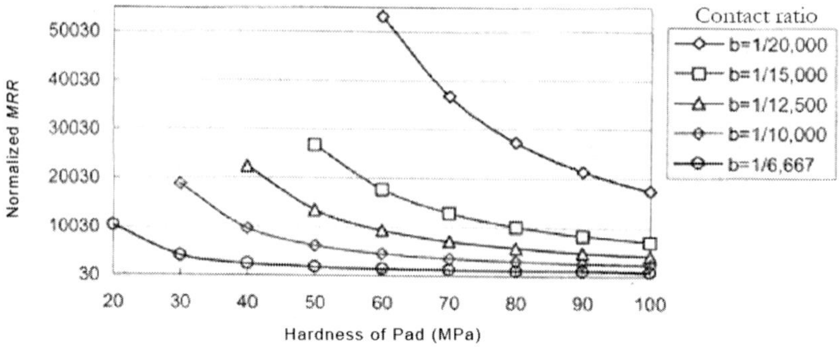

a

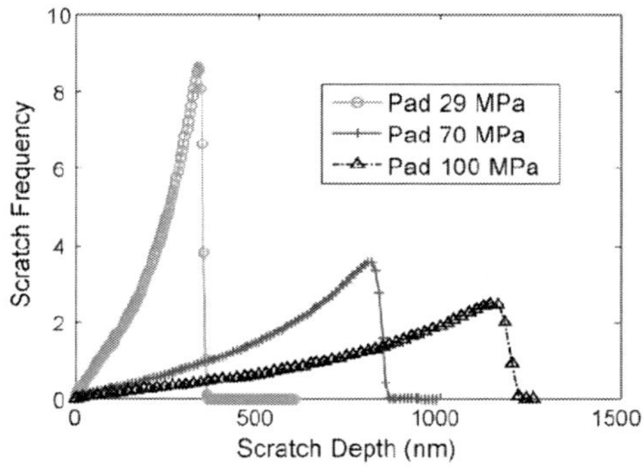

b

Figure 26. Effect of pad hardness on (a) MRR [26]; (b) Scratch propensity [7].

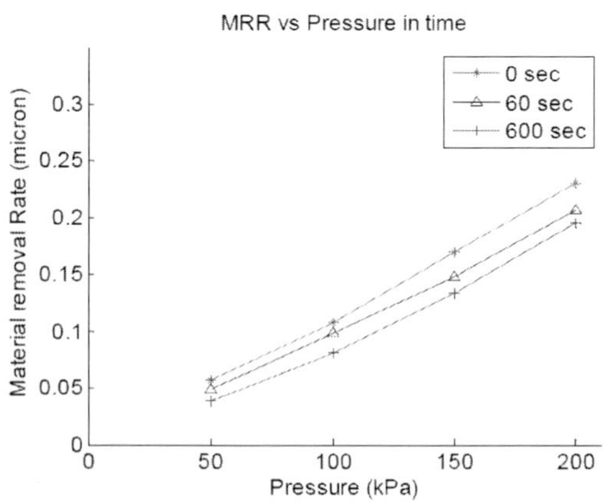

a

Figure 27. (Continued).

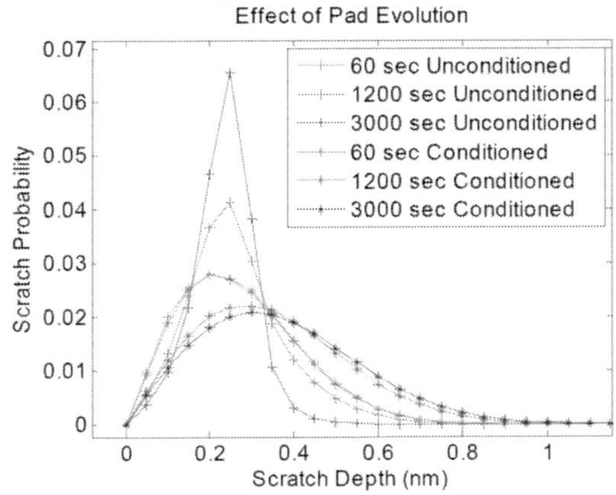

b

Figure 27. Effect of pad conditioning on (a) MRR (no conditioning performed) [25]; (b) Scratch propensity [25].

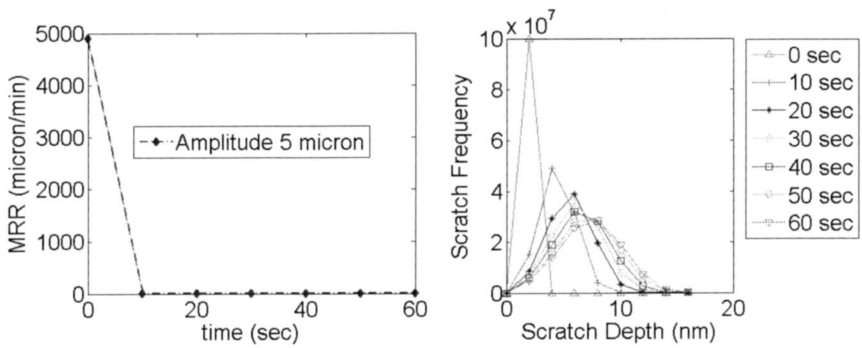

a b

Figure 28. Effect of DLA of silica particles on (a) MRR; (b) Scratch propensity.

For the polishing pad, the parameters are hardness and whether conditioning is performed. For the chemical slurry and abrasive particles, the parameter is pH, which strongly influences whether RLA or DLA applies. The simulation results appear next, followed by a discussion of the results.

In the scratch propensity plots which have the vertical axis labeled as "Scratch Frequency", the vertical axis is equivalent to "Scratch Probability" because frequency is transformed into probability through a straightforward normalization by the total number of scratches occurring in the simulation.

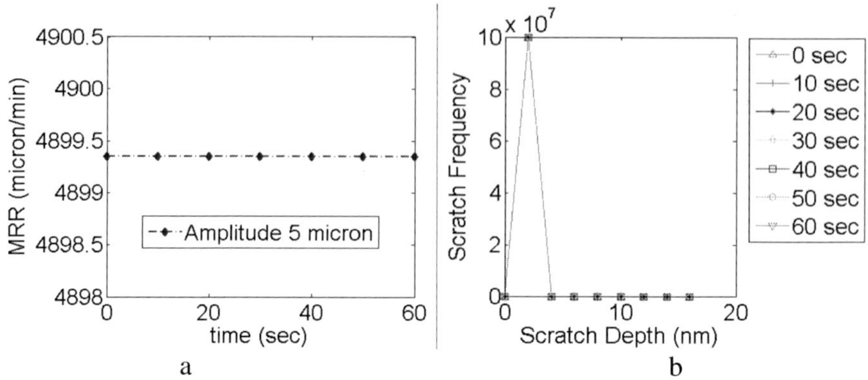

Figure 29. Effect of RLA of silica particles on (a) MRR; (b) Scratch propensity.

Discussion of Results and Conclusion

It is important to understand how MRR and scratch propensity are affected by pad hardness, pad conditioning, and slurry particle agglomeration.

The hardness of the polishing pad has an inverse relationship with MRR [26, 7]. Two hypothesized mechanisms are at work. First, as the pad's hardness increases, fewer abrasive particles can be trapped or embedded into the pad's surface, for example, by the pad's asperities. This means that fewer "active" particles are at work in performing "two-body abrasion"; since MRR is proportional to the number of active abrasives, MRR declines as pad hardness rises [26].

Second, as the pad's hardness increases, there is less direct pad-workpiece contact; contact with the workpiece occurs through more particles that are "inactive" (i.e. performing "three-body abrasion" which is much less effective at removing material than "two-body abrasion") or unconstrained (i.e. not plowing due to entrapment or embedding in the pad). With more particles transmitting force from the pad to the workpiece (but fewer active particles), the decreased force per particle lowers the MRR. The result of these two mechanisms in diminishing MRR with increasing pad hardness is evident in Figure 26a. This, of course, only holds so long as the pad remains relatively soft compared to the workpiece material. Thus, the pad by itself is unable to remove workpiece material. The simulation results from CMPSim in Figure 26b show that as pad hardness (which has a direct relationship with Young's modulus) increases, scratch frequency decreases, but the average scratch depth worsens (increases).

Therefore, with increasing pad hardness, the CMP process produces fewer but much deeper scratches. Thus, both MRR decreases as well as defectivity increases. A soft pad gives good MRR with decreased scratch depth. However, it is detrimental for global planarity. Accordingly, the ideal pad should be locally soft but globally hard.

The conditioning status of the polishing pad affects MRR. As reflected in the CMPSim simulation results in Figure 27a, if the pad does not undergo conditioning during CMP, the MRR declines over time, which is consistent with experimental findings [31, 25, 33, 28, 45]. Several effects are evident from the CMPSim results in Figure 27b. First, if the pad is not conditioned during CMP, the scratch propensity worsens; however, with pad conditioning, scratch propensity improves. Second, with or without pad conditioning, the average scratch depth increases slightly. Third, the improvement in scratch propensity between non-conditioned pad and conditioned pad grows over time, despite a minor increase in average scratch depth. Thus, pad conditioning benefits both MRR and scratch propensity.

The slurry particle agglomeration mode has a significant effect on MRR. DLA rapidly coagulates the slurry particles so that the MRR quickly drops to zero, as shown in Figure 28a, whereas the slurry particles remain separated in RLA so that the MRR stays constant, as shown in Figure 29a. In DLA, scratch propensity improves while the average scratch depth grows over time, which is undesirable, as Figure 28b shows, whereas scratch propensity is unchanged from the very desirable state (because all or most of the probability is for shallow scratches to form) in RLA, as Figure 29b shows.

From the results of this brief parametric study, three recommendations can be made regarding decisions for CMP consumables. First, CMP should be performed in the RLA regime instead of the DLA regime; this offers good MRR and stable, desirable scratch propensity. Second, the pad should be conditioned throughout the CMP process in order to keep the MRR and the scratch propensity steady. Third, the ideal pad should be locally soft but globally hard. Such a pad will be produce good MRR with low defectivity and also be able to maintain good planarity.

Today, several technology industries rely on CMP for uniform flatness at the wafer-scale and precise smoothness at the nano-scale. The important role played by the consumables used in the CMP process requires analytical models to be developed for the polishing pad and the slurry abrasive particles that address MRR and scratch generation propensity. An understanding of the effects of consumables leads to three recommendations for improving material removal and scratch performance.

First, the slurry pH should be chosen such that RLA is occurring. Second, the pad should be conditioned throughout CMP. Third, the pad should be locally soft but globally hard in the CMP process. There are many opportunities for exploration and improvement in CMP, aided by model-based computational simulation software such as CMPSim, with the goal of maximizing yield and defectivity performance.

ACKNOWLEDGMENTS

This work is supported by NSF through grant numbers CMMI 0900093 and CMMI 1100066. The authors gratefully acknowledge this support. Any opinions, conclusions, or recommendations expressed are those of the authors and do not necessarily reflect the views of the sponsoring agencies.

REFERENCES

[1] Archard, J. F. (1953). Contact and Rubbing of Flat Surfaces. *Journal of Applied Physics,* Vol. 24, No. 8, 981-988.

[2] Bastawros, A. F., Bart-Smith, H., Evans, A. G. (2000). Experimental Analysis of Deformation Mechanisms in a Closed-Cell Aluminum Alloy Foam. *Journal of Mechanics and Physics of Solids*, Vol. 48, No. 2, 301-322.

[3] Bastawros, A. F., Chandra, A., Guo, Y., Yan, B. (2002). Pad Effects on Material-Removal Rate in Chemical-Mechanical Planarization. *Journal of Electronic Materials,* Vol. 31, No. 10, 1022-1031.

[4] Biswas, R., Han, Y., Karra, P., Sherman, P., Chandra, A. (2008). Diffusion-Limited Agglomeration and Defect Generation During Chemical Mechanical Planarization. *Journal of the Electrochemical Society,* Vol. 155, No. 8, pp. D534-D537.

[5] Borucki, L., (2002). Mathematical Modeling of Polish-Rate Decay in Chemical-Mechanical Polishing. *Journal of Engineering Mathematics*, Vol. 43, 105-114.

[6] Chandra, A., Bastawros, A. F., Wang, X. P., Karra, P., Haugen, M. (2012). An Integrated Wafer Surface Evolution Model for Chemical Mechanical Planarization (CMP). *Micro-Manufacturing*, Jain, V. K. (editor), Hoboken, CRC Press, 373-400.

[7] Chandra, A., Karra, P., Bastawros, A. F., Biswas, R., Sherman, P. J., Armini, S., Lucca, D. A. (2008). Prediction of Scratch Generation in Chemical Mechanical Planarization. *CIRP Annals – Manufacturing Technology,* Vol. 57, 559-562.

[8] Che, W., Bastawros, A. F., Chandra, A. (2006). Surface Evolution During the Chemical Mechanical Planarization of Copper. A*nnals of the CIRP*, Vol. 55, No. 1.

[9] Che, W. (2005). Synergy Between Chemical Dissolution and Mechanical Abrasion During Chemical Mechanical Polishing of Copper. *Ph.D. Dissertation*, Iowa State University, UMI 3172205.

[10] Che, W., Guo, Y., Bastawros, A. F., Chandra, A. (2002). Mechanistic Understanding of Material Detachment During CMP Processing. *MRS Proceedings,* Vol. 732, 2002 Materials Research Society Spring Meeting.

[11] Che, W., Guo, Y., Chandra, A., Bastawros, A. F. (2004). A Scratch Intersection Model of Material Removal During Chemical Mechanical Planarization (CMP). *Journal of Manufacturing Science and Engineering,* Vol. 127, No. 3, 545-554.

[12] Eusner, T., Saka, N., Chun, J.-H., Armini, S., Moinpour, M., Fischer, P. (2009). Controlling Scratching in Cu Chemical Mechanical Planarization. *Journal of the Electrochemical Society*, Vol. 156, No. 7, pp. H528-H534.

[13] Evans, C. J., Parks, R. E., Roderick, McGlauflin, M. L. (1998). Rapidly Renewable Lap: Theory and Practice. *Annals of the CIRP*, Vol. 47, No. 1, 239-244.

[14] Fu, G., Chandra, A. (2002a). A Model for Wafer Scale Variation in Material Removal Rate in Chemical Mechanical Polishing Based on Viscoelastic Pad Deformation. *Journal of Electronic Materials*, Vol. 31, No. 10, 1066-1073.

[15] Fu, G. (2002b). Modeling of Chemical Mechanical Polishing at Multiple Scales. *Ph.D. Dissertation*, Iowa State University, UMI 3061830.

[16] Fu, G., Chandra, A. (2003). An Analytical Dishing and Step Height Reduction Model for Chemical Mechanical Planarization (CMP). *IEEE Transactions on Semiconductor Manufacturing,* Vol. 16, No. 3, 477-485.

[17] Fu, G., Chandra, A., Guha, S., Subhash, G. (2001). A Plasticity-Based Model of Material Removal in Chemical-Mechanical Polishing (CMP). *IEEE Transactions on Semiconductor Manufacturing*, Vol. 14, No. 4, 406-417.

[18] Gibson, L. J., Ashby, M. F. (1997). *Cellular Solids,* Second Edition. Cambridge, Cambridge University Press.

[19] Greenwood, J. A., Tripp, J. H. (1967). Elastic Contact of Rough Spheres. *Journal of Applied Mechanics*, Vol. 34, No. 1, 153-159.

[20] Greenwood, J. A., Williamson, J. B. P. (1966). Contact of Nominally Flat Surfaces. *Proceedings of the Royal Society of London Series A, Mathematical and Physical Sciences*, Vol. 295, No. 1442, 300-319.

[21] Guo, Y., Chandra, A., Bastawros, A. F. (2004). Analytical Dishing and Step Height Reduction Model for Chemical Mechanical Planarization (CMP) with a Viscoelastic Pad. *Journal of the Electrochemical Society*, Vol. 151, No. 9, G583-G589.

[22] Han, Y. (2006). Scratch Generation Probability in Chemical Mechanical Polishing (CMP) Process. *Masters Thesis*, Iowa State University.

[23] Izumitani, T. (1979). Polishing Lapping and Diamond Grinding of Optical Glasses. *Treatise on Materials Science and Technology*, Tomozawa, M. and Doremus, R. H. (editors). New York, Academic, Vol. 17, 115-172.

[24] Johnson, K. L. (1985). *Contact Mechanics*, Cambridge, Cambridge University Press.

[25] Karra, P. (2009). Modeling and Control of Material Removal and Defectivity in Chemical Mechanical Planarization. *Ph.D. Dissertation*, Iowa State University, UMI 3355516.

[26] Luo, J., Dornfeld, D. (2001). Material Removal Mechanism in Chemical Mechanical Polishing: Theory and Modeling. *IEEE Transactions on Semiconductor Manufacturing,* Vol. 14, No. 2, 112-133.

[27] Luo, J., Dornfeld, D. (2004). *Integrated Modeling of Chemical Mechanical Planarization for Sub-Micron IC Fabrication*, Berlin, Springer-Verlag.

[28] Luo, J., Dornfeld, D. (2003). Effects of Abrasive Size Distribution in Chemical Mechanical Planarization: Modeling and Verification. *IEEE Transactions on Semiconductor Manufacturing,* Vol. 16, No. 3, 469-476.

[29] Luo, Q., Campbell, D. R., Babu, S. V. (1997). Chemical-Mechanical Polishing of Copper in Alkaline Media. *Thin Solid Films*, Vol. 311, 177-182.

[30] Luo, Q., Ramarajan, S., Babu, S. V. (1998). Modification of the Preston Equation for the Chemical-Mechanical Polishing of Copper. *Thin Solid Films*, Vol. 335, 160-167.

[31] McGrath, J., Davis, C. (2004). Polishing Pad Surface Characterisation in Chemical Mechanical Planarisation. *Journal of Materials Processing Technology,* Vol. 153-154, 666-673.

[32] Nolan, L., Cadien, K. (2012). Chemical Mechanical Polish for Nanotechnology. *Nanofabrication*, Stepanova, M. and Dew, S. (editors), Wien, Springer-Verlag, 239-274.

[33] Oliver, M. R. (editor) (2004). *Chemical-Mechanical Planarization of Semiconductor Materials*. Berlin, Springer-Verlag.

[34] Ouma, D. O. (1998). Modeling of Chemical Mechanical Polishing for Dielectric Planarization. *Ph.D. Thesis*, Department of Electrical Engineering and Computer Science, Massachusetts Institute of Technology.

[35] Pandija, S., Roy, D., Babu, S. V. (2007). Chemical Mechanical Planarization of Copper Using Abrasive-Free Solutions of Oxalic Acid and Hydrogen Peroxide. *Materials Chemistry and Physics*, Vol. 102, 144-151.

[36] Prasad, Y. H., Ramanathan, S. (2007). Chemical Mechanical Planarization of Copper in Alkaline Slurry with Uric Acid as Inhibitor. *Electrochimica Acta*, Vol. 52, 6353-6358.

[37] Preston, F. W. (1927). The Theory and Design of Plate Glass Polishing Machines. *Journal of the Society of Glass Technology*, Vol. 11, 214-256.

[38] Ramakrishnan, S., Janjam, S. V. S. B., Patri, U. B., Roy, D., Babu, S. V. (2007). Comparison of Dicarboxylic Acids as Complexing Agents for Abrasive-Free Chemical Mechanical Planarization of Copper. *Microelectronic Engineering*, Vol. 84, 80-86.

[39] Shan, L. (2000). Mechanical Interactions at the Interface of Chemical Mechanical Polishing. *Ph.D. Thesis*, Georgia Institute of Technology, Publication AAI9994449.

[40] Smoluchowski, M. von (1906). Zur Kinetischen Theorie der Brownschen Molekularbewegung und der Suspensionen. *Annalen der Physik,* Vol. 326, No. 14, 756-780.

[41] Steigerwald, J. M., Murarka, S. P., Gutmann, R. J. (1997). *Chemical Mechanical Planarization of Microelectronic Materials*, New York, John Wiley and Sons.

[42] Suphantharida, P., Osseo-Asare, K. (2004). Cerium Oxide Slurries in CMP. Electrophoretic Mobility and Adsorption Investigations of Ceria/ Silicate Interaction. *Journal of the Electrochemical Society*, Vol. 151, No. 10, pp. G658-G662.

[43] Vlassak, J. J. (2004). A Model for Chemical-Mechanical Polishing of a Material Surface Based on Contact Mechanics. *Journal of the Mechanics and Physics of Solids*, Vol. 52, 847-873.

[44] Wang, C., Sherman, P. J., Chandra, A. (2005a). A Stochastic Model for the Effects of Pad Surface Topography Evolution on Material Removal Rate Decay in Chemical-Mechanical Planarization. *IEEE Transactions on Semiconductor Manufacturing,* Vol. 18, No. 4, 695-708.

[45] Wang, C., Sherman, P. J., Chandra, A., Dornfeld, D. (2005b). Pad Surface Roughness and Slurry Particle Size Distribution Effects on Material Removal Rate in Chemical Mechanical Planarization. *CIRP Annals - Manufacturing Technology,* Vol. 54, No. 1, 309-312.

[46] Wang, C. (2004). Stochastic Models for Material Removal Rate (MRR) in Chemical Mechanical Planarization (CMP) Process. *Ph.D. Dissertation,* Iowa State University, UMI 3158376.

[47] Wang, X., Karra, P., Chandra, A., Bastawros, A. F., Biswas, R., Sherman, P., Yao, L. (2007). A Multi-Scale Predictive Model for Wafer Surface Evolution During a CMP Process Incorporating Slurry Evolution. *Twelfth International Conference on Chemical-Mechanical Polish (CMP) Planarization for ULSI Multilevel Interconnection (CMP-MIC),* Fremont, California, March 5-8 2007.

[48] Ye, Y. Y., Biswas, R., Morris, J. R., Bastawros, A. F., Chandra, A. (2003). Molecular Dynamics Simulation of Nanoscale Machining of Copper. *Nanotechnology*, Vol. 14, No. 3, pp. 390-396.

[49] Zantye, P. B., Kumar, A., Sikder, A. K. (2004). Chemical Mechanical Planarization for Microelectronics Applications. *Materials Science and Engineering R*, Vol. 45, 89-220.

In: Recent Advances in Abrasives Research
Editor: Dirk Bähre
ISBN: 978-1-62808-566-2

Chapter 6

FINAL POLISHING OF METALS TO OBTAIN NANOROUGHENED SURFACE WITH NANOPARTICULATE COMPLEX OXIDES OF ALUMINUM AND IRON

L. F. Koroleva
Institute of Engineering Science, Ural Branch,
Russian Academy of Sciences, Ekaterinburg, Russia

ABSTRACT

The concept of the final polishing of hardened steels and nonferrous metals using nanodispersed tribochemical active abrasives based on solid solutions of aluminum and iron oxides is set out in terms of mechanical chemistry. It is demonstrated that the polishing velocity and quality of metals are determined by the formation of an oxide film on their surface and its removal. In order to describe polishing kinetics, the equations for heterogeneous processes are applied. It is shown that the application of these abrasive materials reduces the number of finishing operations with the achievement of a surface roughness no higher than R_a 0.005 μm.

X-ray diffraction, IR spectroscopy, differential thermal analysis, scanning electron microscopy, particle size analysis, and chemical analysis are used to elucidate the general mechanisms of the formation of nanoparticulate solid solutions aluminum oxide and iron oxide prepared via heat treatment of ammonium hydroxycarbonate complexes, $(NH_4)_2Al_2Fe(OH)_5(CO_3)_2nH_2O$. The modification of $Al_{2-x}Fe_xO_3$ and Fe_{2-}

$_yAl_yO_3$ solid solutions oxides (within 0.005 mol. % MoO_2, MnO_2) is shown to enhance the polishing performance of the oxides for final polishing of nonferrous metals and alloys (copper and brass, aluminum) by a factor of 6–7 relative to unmodified aluminum-iron oxides, which is attributable to the increased chemical activity of the abrasive material.

Keywords: Polishing, aluminum oxide, iron oxide, roughness

1. Introduction

The development of precision engineering and electronics requires the processing of metal surfaces to obtain a roughness R_a of less than 0.005 μm. For final polishing (R_a, less than 0.012 μm) abrasive nanodispersed materials are applied with high hardness, for instance, nanodiamond natural and synthesized powders, as well as boron and silicon carbides, cubic boron nitride with grain sizes from 0.01–0.3 μm to 6–10 μm; in addition, ultramicropowders based on aluminum oxide (corundum) and silicon dioxide are also applied [1–11].

Various methods of fabricating nanosized particles of powders based on aluminum oxide are known, for instance, using pulse heating, the deposition of sol-gel of hydroxides of transition metals with subsequent thermal processing, self-propagating high-temperature synthesis (SHS), or mechanical chemistry [12–18].

When these materials are applied, the process of polishing is time consuming and consists of multiple stages. As a rule, the final polishing of precision items is carried out in 3–5 cycles with a consequent decrease in the grain sizes of abrasive material in the composition of pastes and suspensions, which are applied for procedures of primary, final, and fine polishing. Colloidal solutions, for instance, colloidal silicon dioxide, can be applied at the last stages [18].

At the same time, among the most effective abrasive powders are transition-metal and rare-earth oxides with medium hardness (5–7 on Mohs scale): CeO_2, ZrO_2, Cr_2O_3, Al_2O_3, and Fe_2O_3. They are, however, not always capable of ensuring a sufficient abrasion rate or surface roughness class. Oxide solid solutions, e.g., those between chromia-alumina, offer a higher abrasive performance, but fail to ensure surface roughness values below 0.1 μm [19].

The solid state solutions on the basis of oxides chrome and transitive metals have higher abrasive characteristics. Modifying chromium oxide with formation of solid solutions with rare-earth elements and calcium gives an effect of increase in polishing ability and simultaneously provides a roughness of a surface less than R_z 0.10 μm [20-23]. The best solution is, however, to use an environmentally safe abrasive free of chromium.

The choice and working out of new abrasive materials should be based on knowledge of processes of polishing. Development of the theory of polishing is connected with four basic concepts.

The first concept draws an analogy between the phenomena occurring at grinding, and defines process as high-speed mass micro-cutting of surface roughness of firm bodies by a great number of the smallest grinding grains [24]. In the process of metals micro-cutting brought to the local destruction, until education cuts. The polishing is distinguished by the fact that the grain of solid abrasive material, significantly smaller profile unevenness, are so thin scratches, that destruction becomes invisible. But this contradicts the fact of polishing hard materials with softer abrasive powders.

The factors which determine polishing as a mechanical process are as follows: the mean particle size of the largest fraction of abrasive material, the concentration of abrasive powder in the liquid phase, the crystalline state and microhardness of the abrasive and polished metal, the specific pressure on the processed surface, the surface area of contact between the processed surface and the operating surface of the polishing pad, and the polishing velocity. Metal removal C will be determined as follows [24]:

$$C = \frac{c_q v p^r d^u w^f}{S^m} \tag{1}$$

where c_q is the coefficient depending on the properties of material, the abrasive, and processing conditions; v is the polishing velocity; p is the pressure on the processed surface; d is the mean particle size of the largest fraction of abrasive material; w is the weight concentration of abrasive powder in liquid phase; S is the contact area between the processed surface and tool operating surface; and $r = 0.7–1.0$ [25].

Under the second theory the major importance in polishing is given to local plastic deformations. Rebinder come to the conclusion that metal surfaces polishing is the processes when under the influence of external forces in the layers adjoining to a surface there is the flow of metal leading to the message of a maximum smoothness surface [26].

The third concept stated in due time by Beilby [27] as a hypothesis, represents polishing as a melting surface layer and formation of the plasma from the contact points is a result of a friction and a warming owing to energy accumulation. In the process of energy conversion is the highly active similar plasma state high in the free energy of the solid body and briefly-10^{-11}°C temperature can reach high values, for example, up to 10^4-10^5 K and more. This process is application grinding at high speeds. Emergent excited state, called triboplasma, is characterized as severely impaired crystalline structure and presence highly excited crystal fragments and reactionary environment as components of the crystal lattice, ions, and electrons.

The fourth theory of firm bodies polishing which was propose Grebenshchikov [28], Rebinder and Khodakov develop in the sequel [29-32], that connected with sorption mechanochemistry. Under this theory is the formation of the polished surface as result of the frustration of surface layer film by abrasive grains arising in the process of mechanochemical action of the moistening liquid on a surface of a firm body, for example, of glass. Thus, mechanical sorption activation facilitates deformation and mechanical destruction in the adsorbed layer.

Under the concept developed by us, finishing polishing with mechanical and chemical active abrasive dusts is a process of mechanical and chemical activation of a surface of metal and an abrasive material with possible formation of the activated complexes on the metal surface, accompanied: at I stage it is the accelerated oxidation of metal in film nano-layer and oxygen carrying over in film of the nano-layer, at II stage - removal of oxide films by abrasive grains. Thermodynamic formation of the higher oxides with sequence is possible: $Fe/FeO/Fe_3O_4/Fe_2O_3$. Reaction of oxidation of the metal surface at presence of active materials occurs almost instantly to formation oxide films on a surface. Generally, obtaining of the high-quality polished surface of metal is defined with chemical and mechanical factors.

As is obvious from [19–23], high efficient abrasive materials for precision final polishing include abrasives with medium hardness based on oxides of transition metals or their solid solutions. Abrasive material based on the solid solutions of aluminum and iron oxides is environmentally clean; this material is produced by a promising method of synthesizing nanoparticles using the chemical modification of laminar structures applying the formation of hydroxy-complexes $M_{1-x}^{2+}M_x^3(OH)_2\,(X^{n-})_{\,x/n}\,\cdot mH_2O$ [33].

It is also known that the quality of polishing is determined by the structural state and technology of preparation of the processed body surface, the boundary conditions of solids (the value of the surface energy and the composition of adsorbed layers), and the following processes: adsorption, adhesion and the wetting of the surface of the metal and abrasive grain.

Mechanical polishing is a complex interaction between the processed material, the polishing pad, and abrasive particles. When ultrafine nanodiamond, corundum, and boron nitrides are applied, i.e., abrasives with high hardness, polishing is mainly a mechanical process of microcutting roughness and relief smoothing.

The application of nanodispersed tribochemical active abrasive materials significantly modifies the essence of polishing, which is transformed from a mechanical process into a mechanochemical one. As a result, an advantage is obtained over conventional abrasives, since it facilitates a reduction in the amount of finishing procedures; i.e., final and fine polishing can be carried out in one stage with the fabrication of a surface with a roughness of R_a 0.005 μm [33, 34].

A tribochemical process is observed in time with the polishing of sapphires (corundum) by colloidal silicon dioxide with the formation of complex silicates in a surface and their removal. Herewith, the polishing velocity and quality depend on the rate of the chemical reaction of the interaction between sapphire aluminum oxide and abrasive silicon dioxide [35].

In this work we describe in terms of mechanical chemistry the concept of the final polishing of hardened steels with an austenite–martensite structure and nonferrous metals using nanodispersed tribochemical active abrasives. The development of the synthesis principle of nanodispersed tribochemical active abrasive materials based on the solid solutions of transition metal oxides for final polishing is a challenging approach.

The purpose of this work was developing of the mechano-chemical theory of finish polishing and creation of effective abrasive material. The process of polishing austenitic-martensitic steel with abrasive powders based on mixed oxide (oxides solid solution) has been studied from the perspective of mechanochemistry or tribochemistry.

Final polishing is a multicycle process performed by friction under pressure on the surface of a polished body in the presence of polishing paste or suspension with tribochemical active abrasive particles.

In terms of the mechanics of solids, polishing can be referred to as the process of a sliding friction pair. According to available data [36], polishing is considered the destruction of material in the frames of two principally different processes: regular wear and damage. Herewith, the specific features of deformation and destruction of metal can be interpreted in terms of continuum mechanics taking into account the fact that, under conditions of regular wear, structural adaptability of material occurs and all interactions are localized in a thin layer object of secondary structures related with metal oxidation.

Deformation during friction, as was described in [37], progresses in three stages: at the first stage, elastoplastic deformation takes place; plastic destructive deformation takes place at the second stage, and the largest cracks are generated and spread at the third stage, which leads to the destruction of material. The periodicity of structural changes is detected during multicycle loading. In this case it is possible to apply a phenomenological model of the deformation and destruction of metal described in detail elsewhere [38]. The model is based on a scalar concept about faulty rate Ψ, which can be presented as follows:

$$\frac{\partial \psi}{\partial t} = \frac{CH}{(1-\psi_i)^{\beta i}\,\lambda_{pi}} \qquad (2)$$

where $H = (2\xi ij\xi ij)^{1/2}$ is the intensity of shear rates; ξij are the components of shear rate tensor; t is the time; and λpi, βi, C are the parameters depending on the stress strain state and structure of the polished material.

The quality of polishing and, hence, the yield of nondefective items depends not only on the depth and rate of removal of surface layer, but also on the intensity of accumulation of surface damages ψ. Herewith it is considered that the function $\Psi(t)$ has the values $0 \leq \Psi(t) < 1$, provided that there is no macro destruction, for instance, by the detachment of particles of polished material. The value $\Psi = 1$ corresponds to the state of accumulated damages when the material has exhausted its resource and is being destroyed (detachment in our case). It will be consistent to assume that, in the case of nondefective oxide film, $\psi = 0$.

From the point of view of mechanics of finite elements, the stress distribution as a result of chemical mechanical polishing was considered, but without accounting for the influence of the chemical activity of abrasive suspension, and a 2D-axisymmetrical quasistatic model of the process is proposed [39].

In addition, a theory of the molecular-atomic action of abrasive metal on a polished surface is being developed [40, 41].

The concept of the mechanochemical polishing of metals with the formation of oxide films and their removal is highlighted in [33, 41- 43]. It is known that adding reducing agents to suspension efficiently influences the polishing velocity and the achievement of the required surface roughness. For instance, the addition of hydrogen peroxide to a suspension of silicon dioxide during the polishing of copper significantly accelerates the process due to the transition from mechanical to chemomechanical or mechanochemical polishing. Detailed knowledge of polishing processes is of key importance in designing new abrasive materials. As shown earlier, fine polishing is a mechanochemical process in the surface layer of the metal. There is also solid evidence that, to achieve surface roughness values below 0.02μm, the abrasive material must exhibit chemical (tribochemical) activity in the process of friction [43]. Table 1 shows the values R_a corresponding to polishing mechanism of metals. Thus, the application of paste based on complex oxides of aluminum and iron as an abrasive material, as well as the modification of these oxides, significantly changes the final polishing [32, 41]. In this case, the oxidizing capability of abrasive suspension is exhibited, which both reduces the number of finishing procedures and results in a nanoroughened surface. For instance, the complex oxide of aluminum and iron, obtained from hydroxycarbonate complexes, possesses an increased polishing ability for the final polishing of hardened ShKh-15 steel and yields R_a 0.005 μm, which is attributed to tribochemical activity.

Table 1. The ratio of abrasive hardness and microhardness metal and polishing mechanism (*R_a* parameters)

Metals	Abrasive	Microhardness ratio (MPa) Hv abrasive/metals.	R_a, μm	Polishing mechanism
ShKh-15 Steel	Boron carbide	40000/10000=4	0.012	abrasive
ShKh-15 Steel	Corundum Al_2O_3	18000/10000=1.8	0.04	abrasive
ShKh-15 Steel	$Al_{2-x}Fe_xO_3$	1650/10000 = 1.65	0.005	tribochemical interaction
Copper M-1	Minium Fe_3O_4	1100/1300 =0.85	0.02	abrasive
Brass -63	Crocus Fe_2O_3	1100/1500=0.73	0.02	abrasive
Copper M-1	$Al_{2-x}Fe_xO_3$	1650/1100= 1.5	0.006	tribochemical interaction

2. EXPERIMENTAL

In order to simulate the polishing process, we assembled a specialized installation based on the Metapolan-2 vibration device. A polishing pad was made of felt and coated with polishing paste every 15 min; the paste consisted of water and abrasive powder. The process was monitored by variations in the metal removal rate and surface quality in terms of roughness. Figure 1 illustrates the flowchart of specimen polishing: the applied specific pressure was 0.06–0.18 MPa; the rotation rate of the specimens was 85–90 rpm. During the commercial tests, an AShD machine was used for polishing high precision bearing balls with dimensions of 1/32 inch and 1/16 inch (the ball material was ShKh-15 hardened steel). The abrasive properties of specimens in the course of polishing were studied in accordance with regular methods by varying the polishing ability (productivity in mg/cm^2 min):

$$P = \frac{\Delta M}{S \cdot t} \tag{3}$$

where ΔM is the variation of specimen weight; S and t are the polished surface area and polishing duration, respectively. The results obtained have been processed by means of methods of mathematical statistics.

Surface roughness R_a is the arithmetic mean of the absolute values deviations within the profile length and equal

$$R_a = \frac{1}{n}\sum_{i=1}^{n} |y_i| \tag{4}$$

R_a was determined using a Wyko NT-1100 (Veeko) optical profilometer. This is a true interferometric technique in which light reflected off the sample and light reflected from an internal reference mirror interfere. PSI is suitable for measurements smaller than 150 nm in height. In principle, topographical features smaller than 1nm can be measured. For both VSI and PSI, the working distance is 3-10 mm, depending on which objective lens and field-of-view (FOV) lens are used.

The error in the determination of the R_a was equal, on average, to ± 0.0005 μm. Surface roughness R_a was determined using an atomic force microscope (AFM) in combination with a Nanoscan scanning probe microscope operating in hard contact mode. When in contact regime atomic force microscope is analogous to the profilometer.

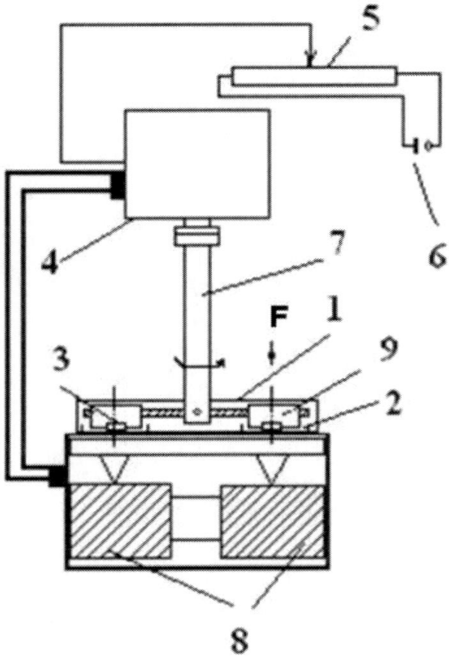

Figure 1. Flowchart of the polishing: 1 – platen; 2 – chute for polishing paste; 3 – specimen; 4 – electric motor; 5 – resistor; 6 – ac source; 7 – spindle; 8 – electromagnets; 9 – load.

Spike the cantilever is in direct contact between the sample and the surface. Is used to determine surface relief with permission from tens of angstroms up to atomic. Scanning is performed using the piezotranslator (Figure 2).

The probe, which is in contact with the surface, point moves along it line by line, and change the X and Y coordinates. In the study of the topography were used of the probes with radii of curvature of the tip from 40 to 10 nm. Topography is registered as a discrete function Z = f (x, y) defined on a two-dimensional array of discrete variables. When measuring the surface roughness at AFM scanning maximum field amounted to 9 x 9 μm with a resolution of 512x512 pixels at a scanning speed 30 μm/s. The error in the determination of the R_a was equal, on average, to ± 0.0002 μm.

Electron microscopy images of the metal surface were acquired using a Tescan VEGA II XMU scanning electron microscope with an INCA WAVE 700 X-ray wave dispersive microanalysis system and an INCA ENERGY 450 energy dispersive microanalysis system with an ADD detector.

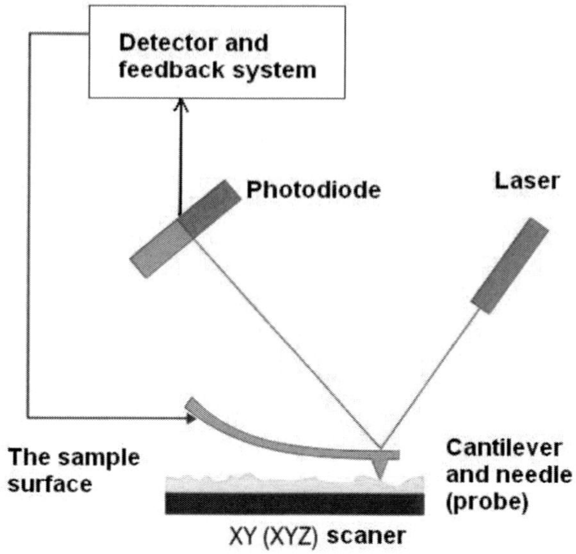

Figure 2. Schema of atomic force microscope (AFM) in combination with a Nanoscan scanning probe microscope operating in hard contact mode.

The microhardness determining produced with relative error of 5%. Hardness testing instrument "Leica" Loading conditions for the ShKh-15: weight = 100 g; time of loading-15s; loading speed – 40μm/s. Loading conditions for copper: mass = 25 g; time of loading-15s; loading speed – 40 μm /s.

Solid solutions of aluminum and iron oxides (complex aluminum and iron oxides) and their modifications were obtained by isothermal processing in the temperature range of 1000–1150°C from ammonium aluminum and iron hydroxycarbonates precipitate in ammonium hydroxycarbonate according to procedures described elsewhere [33, 42]. The abrasive material was modified during the deposition by the addition of molybdate ions (0.1 M solution of $(NH_4)_6Mo_7O_{24}$) and Mn^{2+}. The iron oxide concentration was varied in the range of 0.12–0.175 mol %, the molybdenum dioxide concentration fell in the range of 0.001–0.005 mol %, and the manganese oxide concentration fell in the range of 0.005–0.02 mol %.

The samples were characterized by IR spectroscopy (Shimadzu JR-475 spectrophotometer, KBr disk method) and X-ray diffraction (XRD) (STADI-P diffractometer, nickel-filtered radiation; software for diffraction peak identification using JCPDS–ICDD PDF2 data). Interplanar spacings *d/n* were calculated by the centroid method. The lattice parameters of the complex oxide were determined from the position of the 104, 110, and 113 reflections (uncertainty within 0.003, 0.004, and 0.002 Å, respectively), using the quadratic form for hexagonal symmetry. The volume of the hexagonal unit cell was calculated as $V = \sqrt{3/2}(a^2c)$ with an uncertainty within 0.3 $Å^3$ with differential thermal analysis (DTA) (MOM thermoanalytical system, heating rate of 10–11° C/min, 20–1000°C, 500 mg samples). Particlesize analysis was performed by gravity sedimentation using a Shimadzu SA-CP2 centrifugal analyzer (viscosity of the dispersion medium, 0.0093 P; density of the liquid phase, 1.0 g/cm^3) and by scanning electron microscopy (SEM).

3. Results and Discussion

3.1. Mechanochemical Polishing

Intraprocesses of mechano-chemical polishing are chemical factors as the tribochemical activity of abrasive particle oxides, which will determine the chemical transformations in the near-surface metal layer. In terms of chemistry, the disordering and breakage of chemical bonds is accompanied by a high concentration of active centers on the metal surface. As a result of the force action of abrasive grains, stresses occur on the surface and mechanochemical activation can exist simultaneously on contact points of the metal surface, leading to the initiation of chemical reactions. Under the action of friction in the presence of chemically active abrasive material and mediums, the nearly instant oxidation of the surface layer and the formation of an oxide film, which will be destructed and then removed from the surface, occur. In this case, the oxidizing capability of abrasive suspension is exhibited, which both reduces the number of finishing procedures and results in a nanoroughened surface. For instance, the complex oxide of aluminum and iron, obtained from hydroxycarbonate complexes, possesses an increased polishing ability in 3-5 times for the final polishing of hardened ShKh-15 steel and yields less than R_a 0.005 μm, which is attributed to tribochemical activity [33, 43]. In the course of polishing hardened steel, ShKh-15 oxide film is formed, which is removed in 30 - 60 min (Figure 3). The ShKh-15 steel samples are with an initial surface roughness *Ra* = 0.3–1.3 μm. Embedded abrasive particles are detected on the surface: a chemical analysis of the surface yields a value of 0.05–0.08 wt % Al. The presented data obtained using atomic force microscopy give the following roughness values: R_a less than 0.005 μm. Figure 4 presents hardened ShKh-15 steel samples, polished during 15 min: 2a – with solid solutions of aluminum and iron oxides, 2b – modified molybdenum dioxide solid solutions of aluminum and iron oxides. It is clearly visible on the oxide film surface, with more thick film covering, the entire surface is formed by applying the modified solid solutions of aluminum and iron oxides. The application of solid solutions of transition metal oxides provides the effect of chemical action on the surface, since this system is no equilibrium and chemically active [21, 33, and 42]. This leads to mechanochemical activation of the metal surface and abrasive material and may result in the formation of activated complexes. The oxide films on surface of metals thickness is subdivided into three groups: thin (invisible) which thickness from several angstrom to 40 nm; medium (giving oxide tint) which thickness 40-500 nm; thick visible which thickness of over 500 nm. In the process of polishing with tribo chemical active abrasive in the beginning process are thick films, then the middle and thin, well save engineered surface. The thickness of the oxide film at diffusion-kinetic growth mode is described by an exponential law [44]:

$$h^n = k^n \tau, \qquad (5)$$

where h is the film thickness; τ is the duration of films formation; k, n is coefficients

$$h^2 = k\tau + C, \qquad (6)$$

where k, C is constant.

According to the theory of the growth of thin oxide films Cabrera and Motta [45, 46] controlling phase of metal oxidation was the ion migration is by influence of the accelerating electric field inside a film formed on the boundary of metal-film positive metal ions and negative chemisorbed oxygen ions. Final polishing with complex oxides is based on the oxidation of the surface. An alteration of the electron subsystem, for instance, for steel, occurs: oxidation

$$Fe^0 \rightarrow Fe^{2+} \rightarrow Fe^{3+}.$$

The electron spin changes in partially occupied shells can be presented as follows:

$3d^6 4s^2 \rightarrow 3d^6 4s^o \rightarrow 3d^5 4s^o$

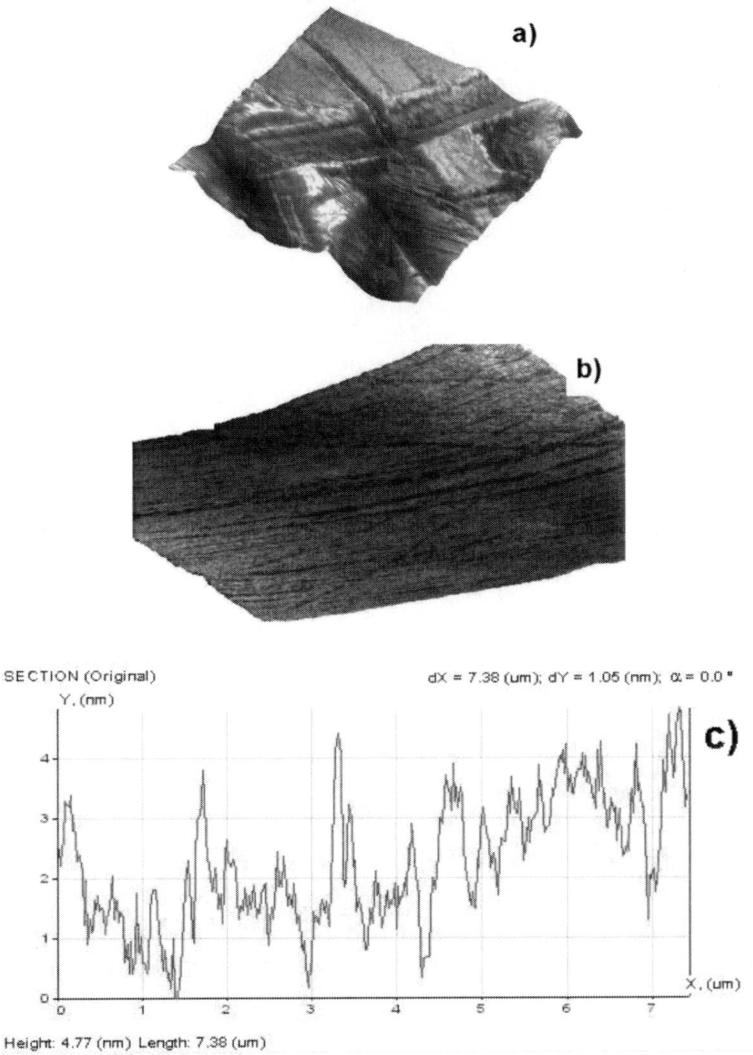

Figure 3. Typical AFM images and surface profiles of ShKh-15 steel specimens: (a) before polishing (Ra = 0.2, μm), ShKh-15 steel polished for 30-60 min (b), characteristic AFM image of surface relief (c), surface cross section (c), Ra 0.004±0,0002 μm.

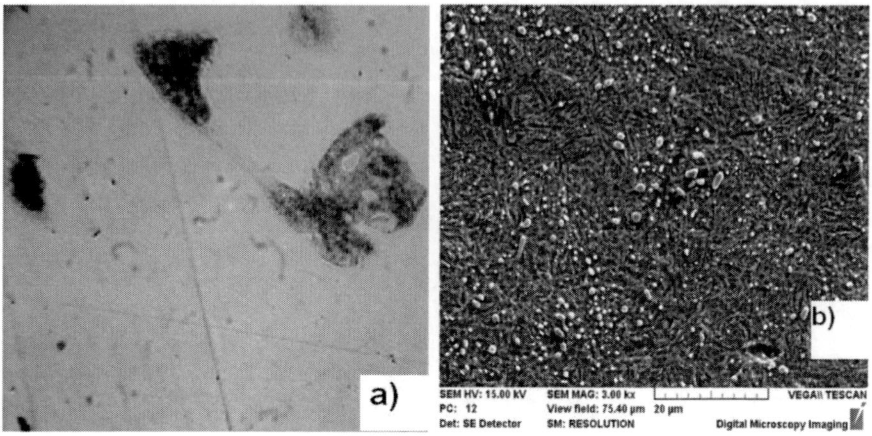

Figure 4. The oxide film surface hardened ShKh-15 steel samples, polished during 15 min: (a) solid solutions of aluminum and iron oxides; (b) modified molybdenum dioxide solid solutions of aluminum and iron oxides.

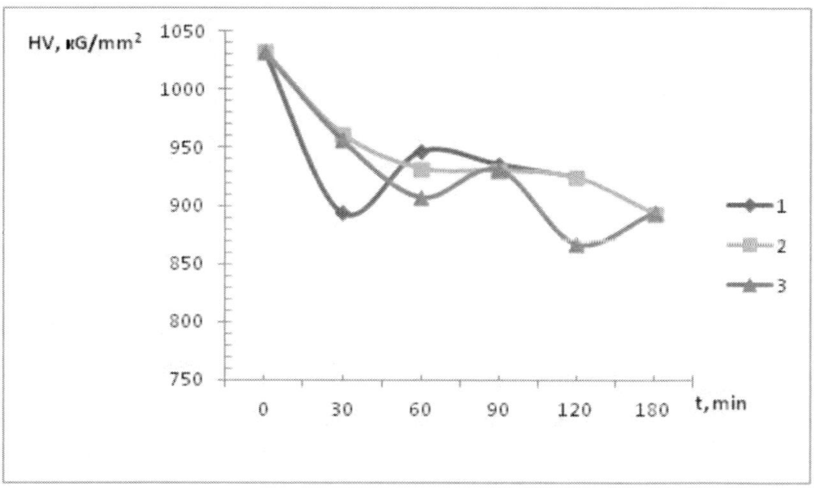

Figure 5. Microhardness of ShKh-15 steel as a function of polishing duration: (1) with modified chromium oxide; (2) with complex oxides of aluminum and iron; (3) in vibration mode (3000 rpm). The experimental results obtained have been processed by means of methods of mathematical statistics.

The application of a vibration mode (3000 rpm) accelerates polishing, but the surface roughness R_a increases to 0.060 μm in comparison with the mode without vibration.

The variation of the microhardness of the specimens is insignificant: from 1050 kgf/mm^2 to 920 kgf/mm^2 (Figure 5, Experimental data). The hardness of the unpolished tempered steel ShKh-15 is 1000-1050 HV. A slight decrease of microhardness confirms the nature of the mechanochemical polishing. When mechanical polishing only has increases of microhardness.

However, non-ferrous metals and alloys such as copper and brass and those more resistant against oxidation such as titanium are polished less efficiently by a complex oxide of aluminum and iron. In order to increase the yield of polishing of nonferrous metals and alloys, it is necessary to increase the oxidizing capacity of abrasive material or suspension by the addition of an oxidizing agent. Modification by molybdenum or manganese oxide, which is characterized by the ability to oxidize and form compounds with varying valence and, hence, to be tribochemical active, provides such effect. It is established that modification by molybdenum and manganese oxides in the amount of 0.001–0.01 mol. % MoO_2 significantly increases the polishing efficiency [42].

Specimens of aluminum alloy D-16 with nanoroughened surfaces were fabricated using a complex modified oxide of aluminum and iron for polishing. In the presented electron microscopy images of the alloy D-16 surface, it can be seen that, at the start of the process, the oxide film is destructed (light particles) and then removed (Figure 6). The existence of parallel grooves on the metal surface as a result of plastic deformation should be mentioned. When copper and its alloys are polished, a complex oxide of aluminum and iron modified by molybdenum oxide (0.001–0.01 mol % MoO_2) is most efficient.

Figure 7 depicts a copper specimen and yellow metal obtained by polishing: parallel grooves after plastic deformation are visible. R_a determined by AFM, The experimental results obtained have been processed by means of methods of mathematical statistics.

For illustration see the characteristic AFM image of surface relief. The calculation is made with the same base length.

Therefore, it can be concluded that, upon the final polishing of metals such as copper and aluminum, its alloys, both the chemical oxidation of the surface and the plastic deformation of metal occur. Final polishing with tribochemical active abrasives is a process where mechanical action initiates chemical transformations leading to surface smoothing; therefore, it can be considered a mechanochemical process. On the basis of investigations into oxide films that formed during polishing, the final polishing with complex oxides can be reasonably considered a mechanochemical process [34, 43].

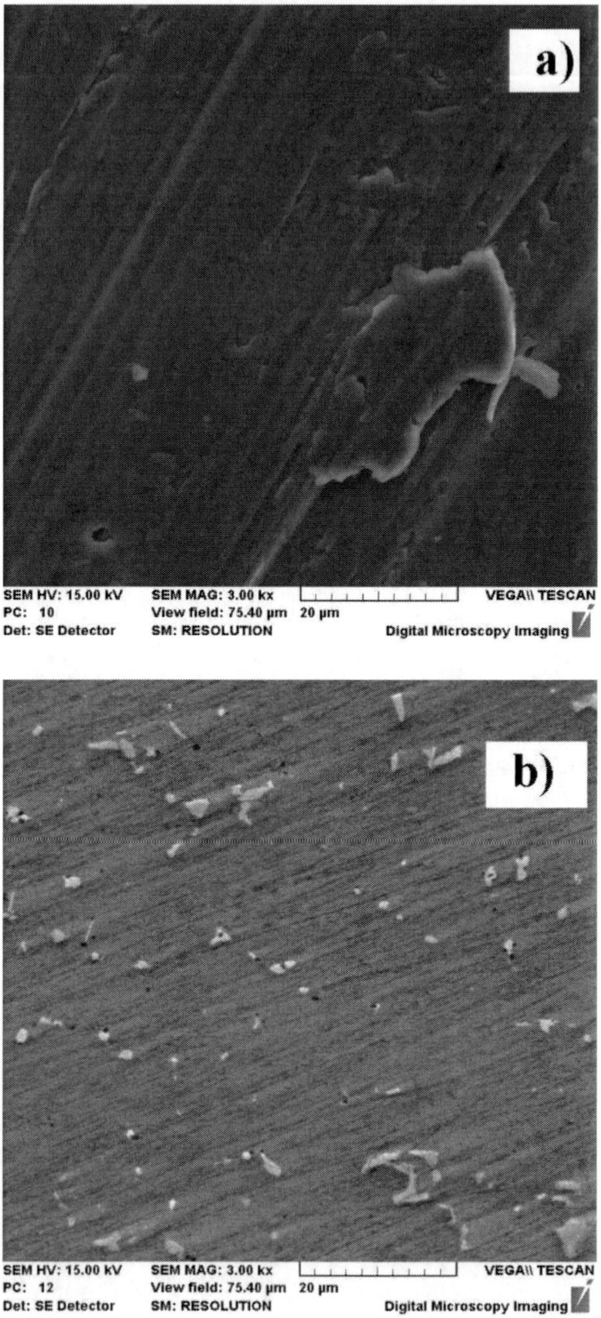

Figure 6. (Continues).

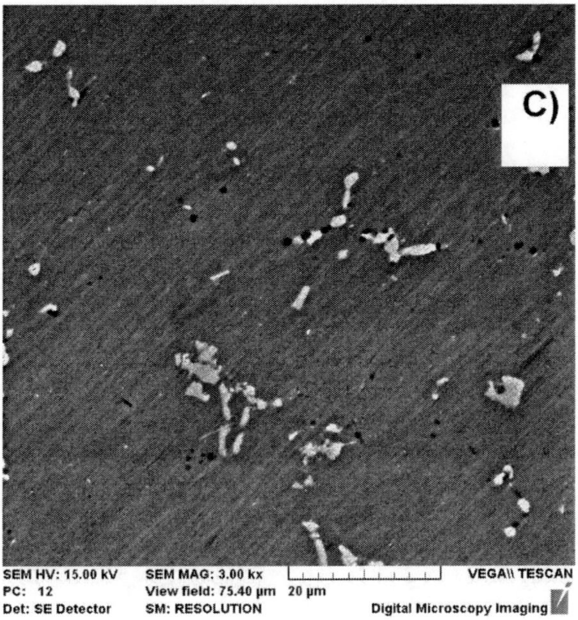

Figure 6. SEM-images of the surface of aluminum alloy D-16: (a) initial; (b) polished for 30 min; (c) polished for 1 h.

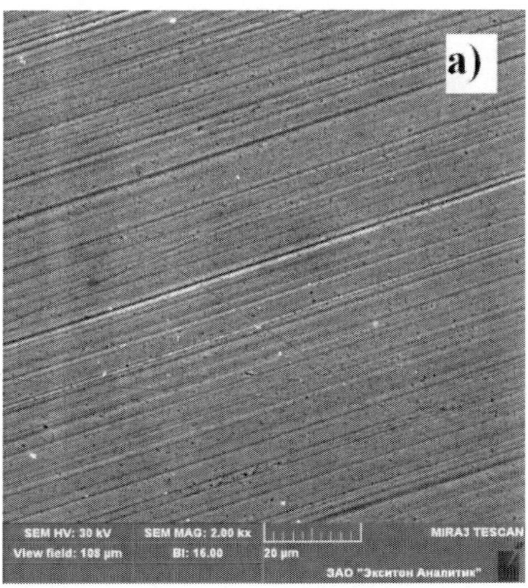

Figure 7. (Continues).

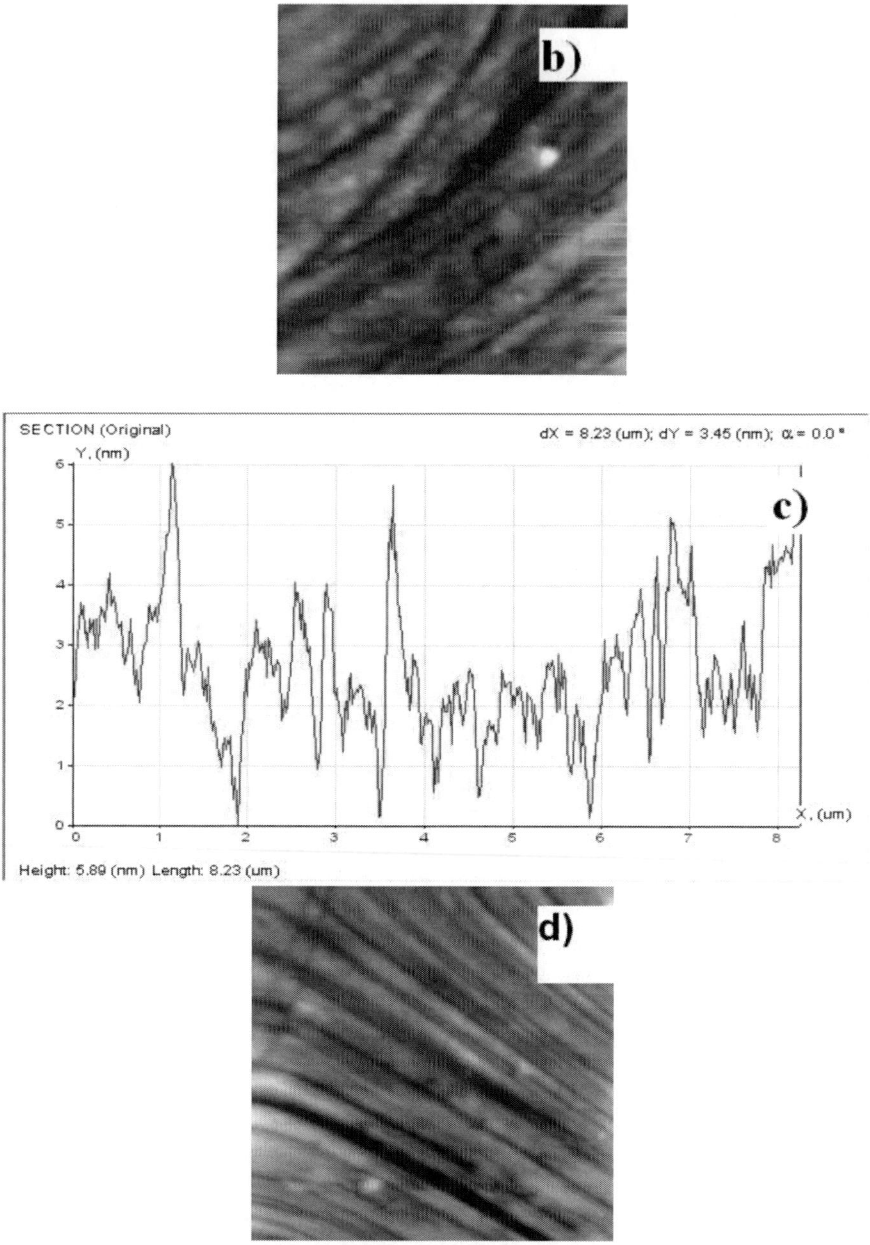

Figure 7. SEM-image of polished surface of copper M-1 (a), characteristic AFM image of surface relief (b), surface cross section: R_a 0.006±0.0003 µm (c), characteristic AFM image of polished surface relief of yellow metal (d).

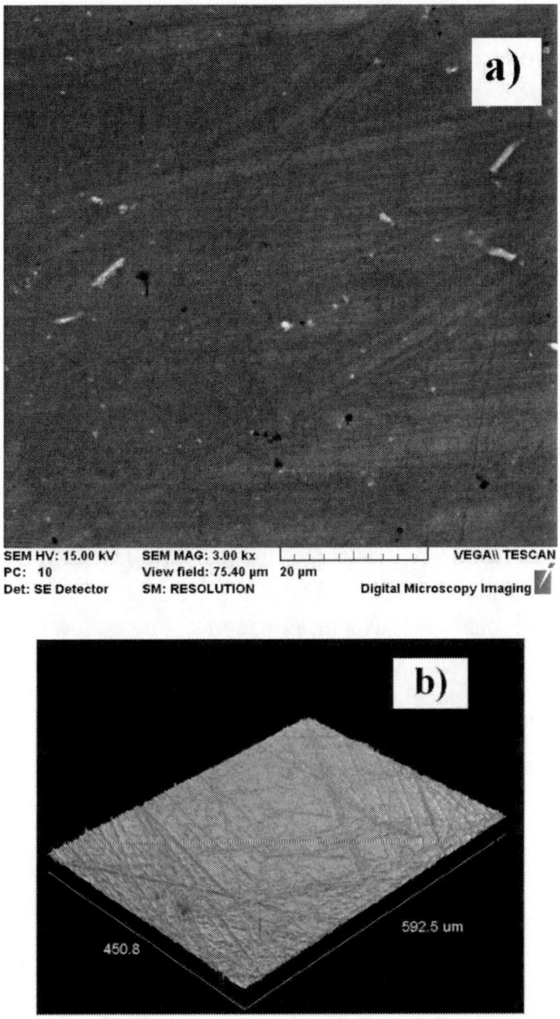

Figure 8. SEM-image (a) of the surface of ShKh-15 steel polished by boron carbide; (b) surface topography: R_a 0.119±0,005 μm.

As was established by our experiment, during polishing with solid abrasive materials such as boron carbide with the identical dispersity as complex oxides of aluminum and iron, the minimum achievable surface roughness R_a is 0.075–0.119 μm (Figure 8) [43].

As a result, under friction in the presence of chemically active abrasive material and mediums, the nearly instant oxidation of the surface layer and the formation of an oxide film, which is destroyed and removed from the surface, occur. The highest degree of oxidation is possible in contact points where micro cracks are generated. As a result of oxidation and the destruction and removal of film, the surface is smoothened. At the initial polishing stage, as a result of the chemical action of paste components and the mechanical action of abrasive particles, oxides are formed in the nearsurface layer, for example, according to the rule proposed elsewhere [47]; in the presence of water they can transfer to hydroxy complexes:

$$Fe + O_2 \rightarrow FeO \rightarrow Fe_2O_3$$

$$Fe_2O_3 + 2H_2O \rightarrow 2FeOOH$$

If mechanochemical polishing is considered a process at the interface between solids (including the oxidation and formation of an oxide phase in the subsurface layer), then the overall system can reasonably be considered a two - phase system with separating surface and ion exchange. Such processes are identified as heterogeneous topochemical; their kinetic regularities can be described by the Erofeev–Kolmogorov–Avrami equations [48]. These equations have been used successfully to describe the kinetics of polishing ShKh-15 steel with the participation of the reactions in the surface layer:

$$\alpha = 1 - e^{-kt^n} \tag{7}$$

$$K_{ck} = nk^{1/n} \tag{8}$$

where α is the degree of transformation of metal atoms to oxides defined as the relative intensity of metal removal $C_i/C_{\max}$ (it is assumed a priori that the factor of the mechanochemical action of the abrasive is the main one), n is the number of consecutive stages during the formation of the center of the new oxide phase, k is the process constant, t is the duration of polishing, and K_{ck} is the constant of the oxidation rate. As a result of the studies of the kinetic regularities of polishing using equation (4) and (5), experimental values n, k, and K_{ck} for polishing nonferrous metals are obtained (Table 2). It is revealed that the lowest value of surface roughness, R_a 0.005 μm, exists at K_{ck} equal to 0.03–0.06. It is established that, for polishing nonferrous metals and alloys such as copper, brass, and aluminum, the modified complex oxide of aluminum and iron (0.001–0.01 mol % MoO_2, Mn_2O_3) is the most efficient.

Table 2. Experimental parameters in kinetic equations of polishing

Alloys	Abrasive material	$R_{a,}$ µm	Parameters of kinetic equation			Standard deviation *
			n	k	$K_{ск}$	S^2
ShKh-15 Steel	Complex oxide of aluminum and iron	≤0.005	1.0	0.04	0.03	0.005
ShKh-15 Steel	Complex oxide of aluminum and iron modified by molybdenum	0.005	1.54	0.005	0.05	0.003
ShKh-15 Steel	Boron carbide B_4C	0.12	1.3	0.01	0.004	0.005
Aluminum alloy D-16	Complex oxide of aluminum and iron	0.02	1.2	0.03	0.06	0.004
Aluminum alloy D-16	Complex oxide of aluminum and iron modified by manganese	0.02	1.0	0.05	0.04	0.005
Copper M-1	Complex oxide of aluminum and iron	0.050	1.2	0.01	0.02	0.003
Copper M-1	Complex oxide of aluminum and iron modified by molybdenum	0.006	1.2	0.03	0.05	0.005

*Standard deviation for determining the coefficients with averaging over five parallel experiments.

The results obtained have been processed by means of methods of mathematical statistics. Figure 8 illustrates kinetic curves of polishing hardened ShKh-15 steel and M-1 copper using various abrasive materials. The existence of MoO_2 in the specimens in amounts of 0.001–0.005 mol % increases the polishing ability by an order of magnitude for the polishing of copper and brass, which is reflected in the kinetic curves. Figure 9 illustrates the kinetic curves of alterations of surface roughness when polishing hardened steel and copper using various abrasives. Increased surface roughness (curve 4) sample ShKh -15 steel by polishing with boron carbide due to mechanical impact abrasive particles.

It should be noted that the effectiveness of the manganese oxide modification of the solid solution of iron and aluminum oxides for polishing of the copper compared with molybdenum dioxide has less productivity.

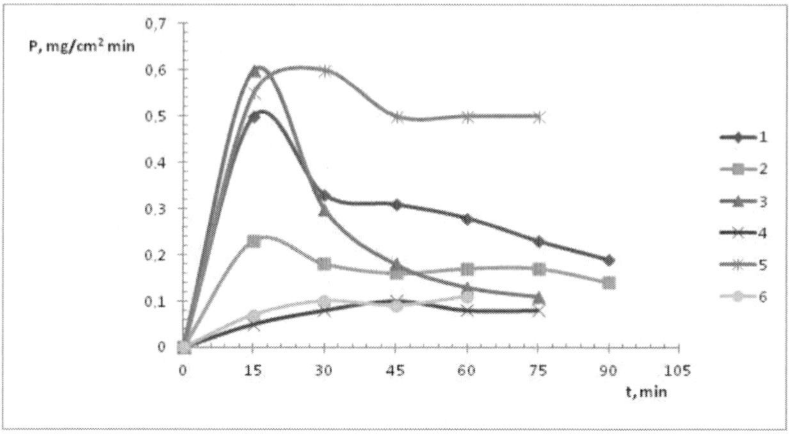

Figure 9. Kinetic curves of polishing: (1, 2) ShKh-15 steel by a complex oxide of aluminum and iron ((1) 0.175; (2) 0.14 mol % Fe_2O_3); (3) ShKh-15 steel by boron oxide; (4) M-1 copper by a complex oxide of aluminum and iron; (5) M-1 copper by a complex oxide of aluminum and iron modified with MoO_2; (6) ShKh-15 steel by a complex oxide of aluminum and iron modified with Mn_2O_3. The experimental results obtained have been processed by means of methods of mathematical statistics.

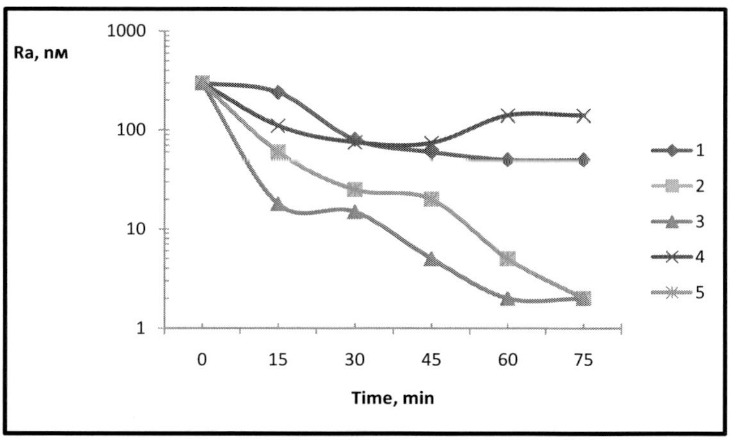

Figure 10. Kinetic curves of variation of surface roughness R_a upon the polishing of specimens: (1) M-1 copper by a complex oxide of aluminum and iron (0.125 mol % Fe_2O_3); (2) M-1 copper by a complex oxide of aluminum and iron modified with MoO_2; (3) ShKh-15 steel by complex oxide of aluminum and iron; (4) ShKh-15 steel by boron carbide; (5) M-1 copper by a complex oxide of aluminum and iron modified with Mn_2O_3. The experimental results obtained have been processed by means of methods of mathematical statistics.

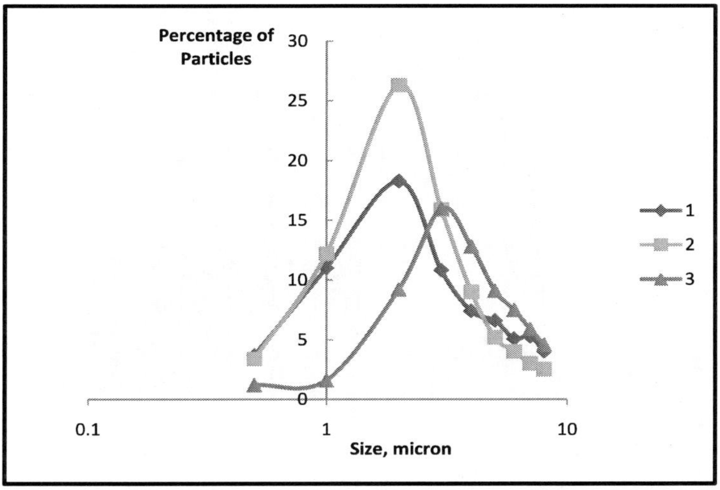

Figure 11. Particle size distributions for an aluminum iron oxide: (*1*) unmodified, (*2*) modified with 0.005 mol % MoO_2, (*3*) modified with 0.001 mol % MoO_2.

This is due to a variety of the electronic state of Mn^{2+} ($3d^5 4s^0$) and Mo^{4+} ($4d^4 5s^0$) and compound: Mn_2O_3, MoO_2, consequently more active oxidant is molybdenum dioxide. While analyzing data on sedimentation and electron microscopy it has been found that specimens of complex oxides of aluminum and iron (Fe_2O_3 0.12–0.156 mol %), which yields a surface roughness of R_a 0.005 μm. Particle-size analysis data show that, in the aluminum iron oxide samples modified with molybdenum dioxide, the fraction of particles 10 nm or smaller in size is the same as in the unmodified samples (4–5%). The main particle size fraction is 50 nm (Figure 10, 11). On the basis of our experimental data, it is possible to conclude that the effect of achieving a minimal surface roughness less than R_a 0.005 μm in one finishing procedure is stipulated by the fact that, during polishing, the abrasive complex oxide of aluminum and iron is tribochemical active. Nanoparticles play the role of an oxidant due to the variation in the oxidation degree of iron, molybdenum, or manganese. The nanoparticles simultaneously act as a lubricant [49]. The role of coarser particles is the force action on the surface relief and the removal of the oxide film with surface smoothing. During the commercial final polishing of high-precision bearing balls made of ShKh-15 hardened steel, which are used in rocket systems, the yield of nondefective items was 85–90%. The application of known abrasive materials (based on corundum and oxides of transition metals) yields less than 50% of nondefective items of the same precision.

3.2. Synthesis of Tribochemical Active Abrasive Materials

One of the promising methods for obtaining nanoparticles includes the chemical modification method of stratified structures that use the education hydroxy-complex $M^{2+}_{1-x}M^{3}_{x}(OH)_2\,(X^{n-})_{\frac{x}{n}} \cdot mH_2O$.

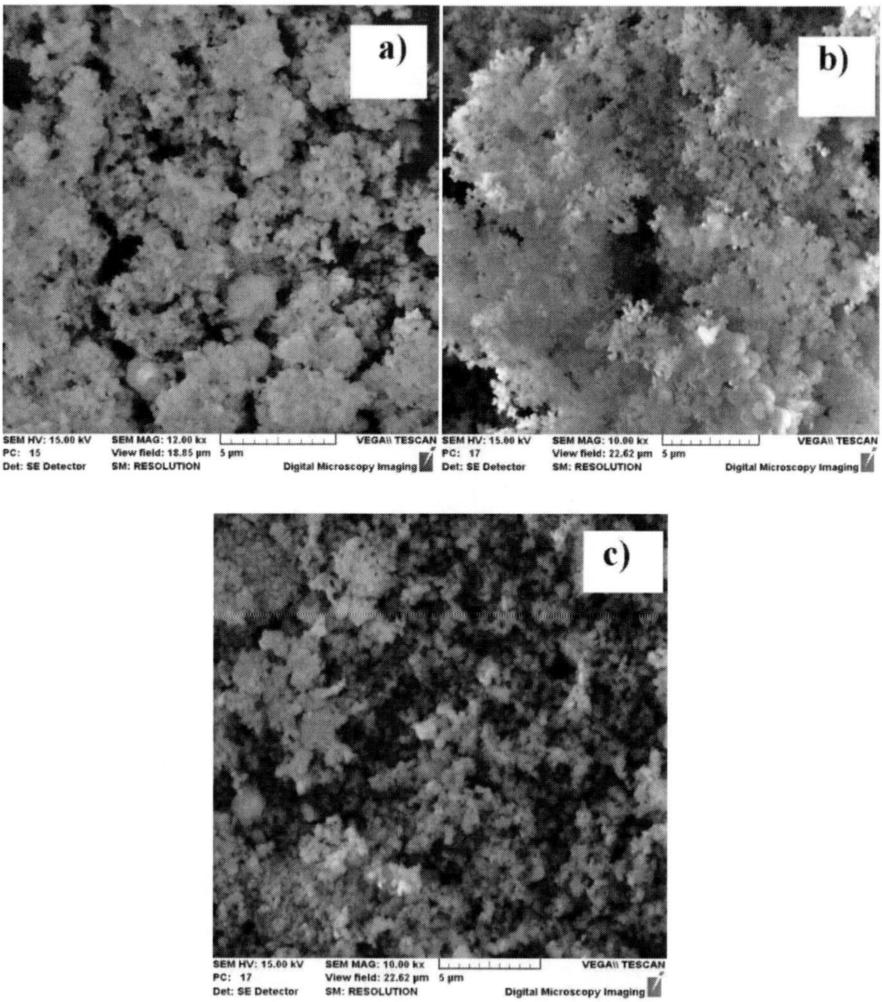

Figure 12. SEM-images of abrasive material based on (a) a complex oxide of aluminum and iron; (b) a complex oxide of aluminum and iron modified with molybdenum oxide; (c) a complex oxide of aluminum and iron modified with manganese oxide.

Looking at the features of this method, the structure of these compounds is composed of positively charged hydroxide layers $M_{1-x}^{2+}\ M_x^3(OH)_2^{x+}$ and anions X^{n-}.

In Figure 12 a schematic of such complexes is presented. The synthesis of hydroxycarbonates-complexes has been used to production of the nanodispersed oxides solid solutions. The aggregate of intermediate hydroxycarbonates formation of complex compounds with following hydrolysis and heat treatment gives the effect of getting qualitatively new functional capabilities. The structure of relationships formed at first-stage deposition and formation of the hydroxycarbonate. A significant advantage of this method is the unlimited possibility to vary the ratio of the atomic elements during the deposition. IR spectroscopy and chemical analysis data indicate that the precipitated aluminum iron ammonium hydroxycarbonates contain five hydroxy groups and two carboxy groups per formula unit and can be represented by the formula:

$$NH_4FeAl_2(OH)_5(CO_3)_2 \cdot nH_2O$$

These ammonium hydroxycarbonates are similar in stoichiometry to the cobalt aluminum chromium hydroxycarbonates described previously [50, 51]. The total reaction of aluminum iron hydroxycarbonate precipitation can be represented by the scheme:

$$\begin{gathered}FeSO_4 + Al_2(SO_4)_3 + 9\ NH_4HCO_3 = NH_4Al_2Fe(OH)_5(CO_3)_2 + 4\ (NH_4)_2SO_4 + 2H_2O + \\ +7\ CO_2\end{gathered} \quad (9)$$

Al^{3+} precipitation in the form of hydroxy salts or hydroxides is known to be accompanied by olation: the formation of polymer structures with OH bridges. Based on the theoretical views of olation processes, the tendency of Al^{3+} Fe^{3+} and Fe^{2+} to form polynuclear complexes, and the present IR spectroscopy and chemical analysis data, the general formula of the ammonium hydroxycarbonates can be written in the form:

Examination of the hydroxycarbonate precipitation conditions indicates that, to obtain well-formed hydroxycarbonate precipitates with good filterability, the temperature must be maintained in the range of 60–75°C. A precipitation above 75°C leads to hydroxycarbonate decomposition and, as a consequence, hydroxide formation, as evidenced by the chemical analysis data.

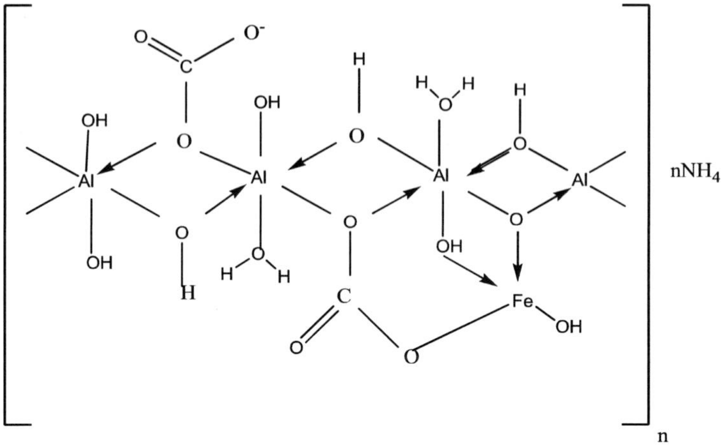

(10)

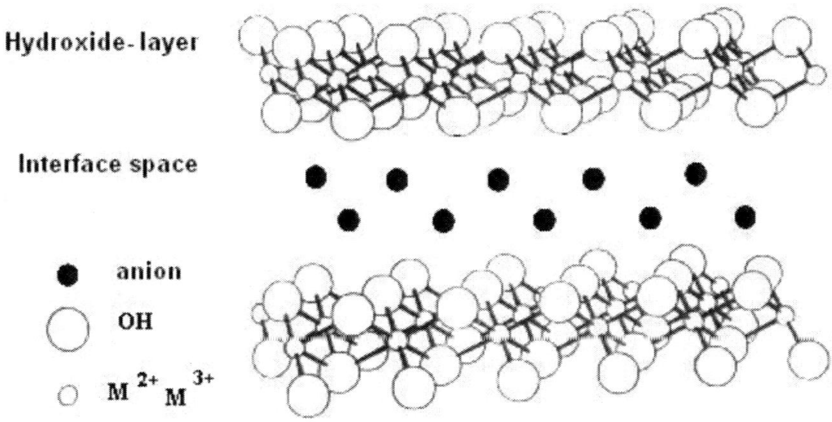

Figure 13. Schematic hydroxy-complex $NH_4FeAl_2(OH)_5(CO_3)_2 \cdot nH_2O$.

The samples obtained below 60°C have low polishing performance. Not less important in hydroxycarbonate precipitation is the solution pH. The pH range for the formation of hydroxycarbonate complexes is 7.5–8.5. Below pH 7.5, hydroxides are precipitated. Above pH 8.5, hydroxycarbonates partially decompose to form hydroxides.

The above formula of the metal ammonium hydroxycarbonates is supported by the fact that their IR spectra (Figure 13) show absorption bands characteristic of groups in bidentate coordination (bonded to different metal cations; C–O stretching mode at 735ν_2, C–O bending mode at 1540 cm^{-1}) and monodentate coordination (bonded to cations of one metal; symmetric stretching modes at ν_2 = 850 cm^{-1} and ν_{2a} = 1390 cm^{-1}, asymmetric stretching mode at ν_{3b} = 1445 cm^{-1}). The absorption bands at 985 and 1100 cm^{-1} are attributable to the bending mode of OH groups, including OH bridges [52, 53]. Consequently, the group replaces the OH bridge in a polynuclear chain, as shown in scheme (10). There is also evidence for the presence of water (bending mode at 1620 cm^{-1} and stretching mode at 3480 cm^{-1}). The DTA curves of the metal ammonium hydroxycarbonates (Figure 14) show a broad endotherm at 220–225°C, accompanied by a weight loss (41.2 to 46.8%) and due to the removal of CO_2, NH_3, and water. The exothermic event at 340°C is the formation of aluminum iron oxyhydroxides. Holding at 425°C leads to a complete removal of the OH groups; the associated weight loss is 50.2–52.5%.

The formation of a solid solution of iron oxide in alumina begins at 950–1000°C, as supported by XRD results (Figure 15). Heat treatment of the hydroxycarbonates at temperatures from 950 to 1200°C leads to the oxidation of divalent to trivalent iron, as evidenced by chemical analysis data. The reaction can be represented by the scheme:

$$4NH_4FeAl_2(OH)_5(CO_3)_2+O_2=2Fe_2O_3+4Al_2O_3+4NH_3+8CO_2+12H_2O \quad (11)$$

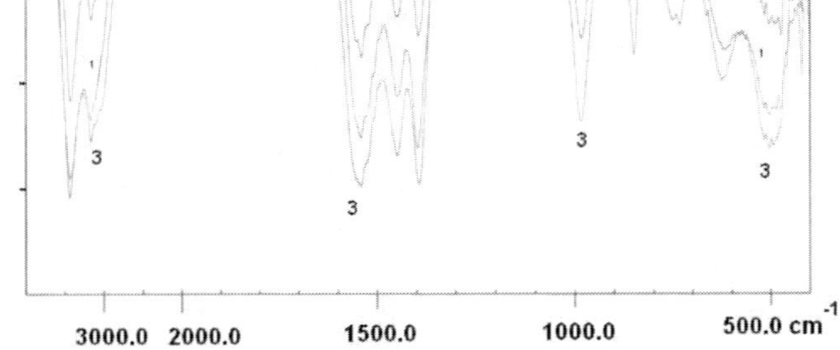

Figure 14. IR-spectra of samples: of (1) ammonium aluminum iron hydroxycarbonates modified with molybdate ions (0.12–0.156 mol % Fe_2O_3, 0.005 mol % MoO_2); (2) ammonium aluminum iron hydroxycarbonates (0.156 mol % Fe_2O_3); (3) ammonium aluminum hydroxycarbonate.

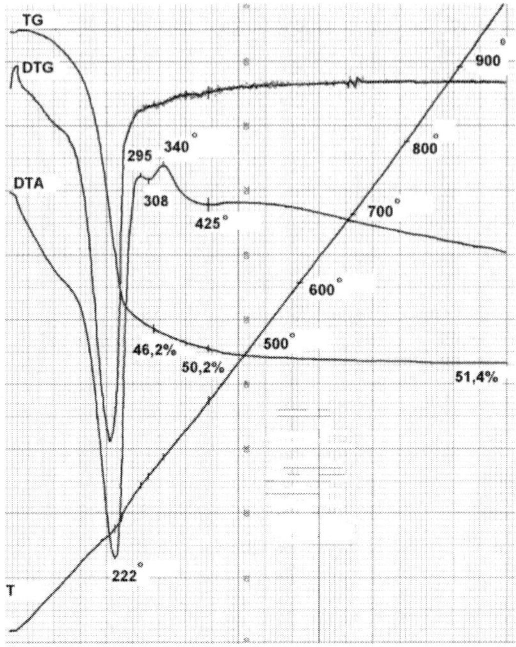

Figure 15. Differential Thermal analysis results for aluminum iron ammonium hydroxycarbonates.

The optimal calcination temperature for achieving high polishing performance is 1150–1200°C. In the composition range $x = 0.03$–0.30, the *a* cell parameter of the corundum-based hexagonal solid solution $\alpha - Al_{2-x}Fe_xO_3$ is proportional to the iron oxide content (Figure 16).

The *c* parameter and the unit-cell volume *V* of the hexagonal solid solutions decrease with increasing iron oxide content. The composition dependences of *a*, *c*, and *V* suggest that the solid solutions obtained in this study differ in properties from those prepared by heat-treating oxide mixtures.

We observe the formation of both corundum- and hematite-based solid solutions, but in small amounts (3–5 wt %), in agreement with previous results for the Fe_2O_3 concentrations under consideration and temperatures below 1300°C [54]. According to earlier data [54, 55], in addition to hematite-based solid solutions the process may lead to the formation of corundum and a spinel phase with iron in the oxidation state 2+. Note that a bonding configuration characteristic of corundum- and hematite-based solid solutions forms during the release of hydroxy and carbonate groups and the formation of mixed oxides from amorphous material.

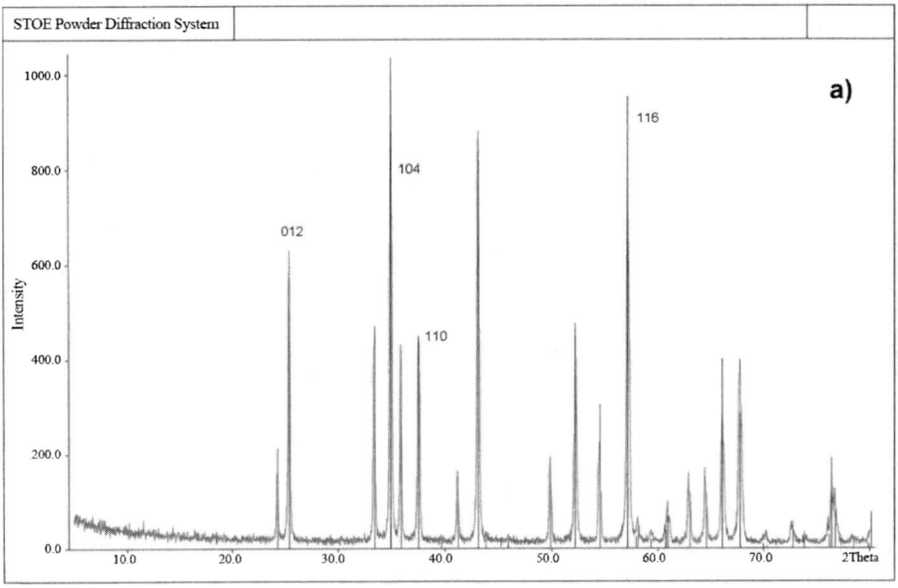

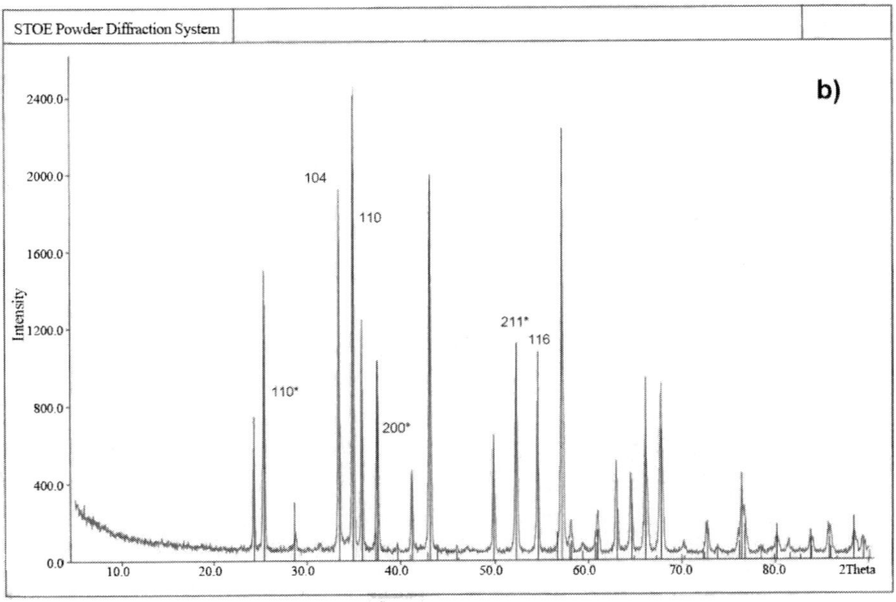

Figure 16. (Continues).

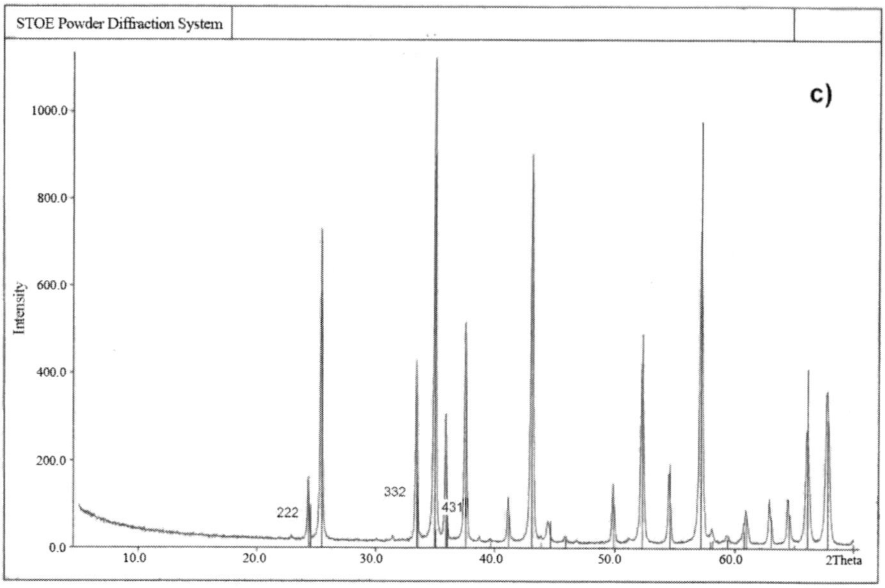

Figure 16. XRD patterns of aluminum iron oxide samples: (a) modified with 0.001 mol % MoO_2 (Miller indices for α-$Al_{2-x}Fe_xO_3$); (b) modified with 0.005 mol % MoO_2 (Miller indices for α-$Fe_{2-y}Al_yO_3$); (c) unmodified (Miller indices for $Al_3Fe_5O_{12}$).

The thermal transformations of the ammonium hydroxycarbonates can be represented by the scheme:

$$NH_4FeAl_2(OH)_5(CO_3)_2\cdot nH_2O \xrightarrow{220\text{-}250^{\circ}C} Al_2Fe(OH)_2O_2 \xrightarrow{425\text{-}450^{\circ}C} Al_2O_3\cdot Fe_2O_3\ _{\text{amorph.}} +$$

$$+\ \alpha-Fe_2O_3 \xrightarrow{950\text{-}1150^{\circ}C} \alpha-Al_{2-x}Fe_xO_3\ _{\text{cryst.}} + \alpha-Fe_{2-y}O_3\ _{\text{cryst.}} \quad (12)$$

The precipitation process was accompanied by the oxidation of Fe^{2+} to Fe^{3+} and the reduction of $(NH_4)_6Mo_7O_{24}$ to $(NH_4)_6Mo_7O_{23}$ in a nonstoichiometric oxidation state (so-called molybdenum blue) [56].

In contrast to molybdate-free aluminum iron hydroxycarbonates, the samples contained Fe^{3+} complexes (up to 1–5 wt %), which resulted from oxidation with molybdate ions. Chemical analysis showed that only one-fifth of the molybdate ions were precipitated, and the rest remained in solution, in accordance with previous reports on incomplete precipitation of molybdate ions in the presence of ammonium.

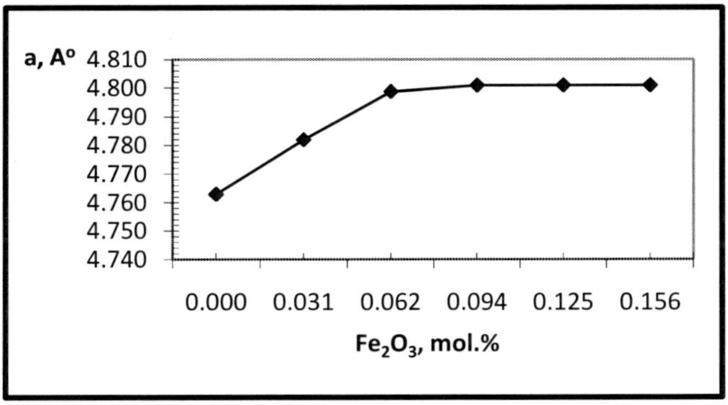

(a)

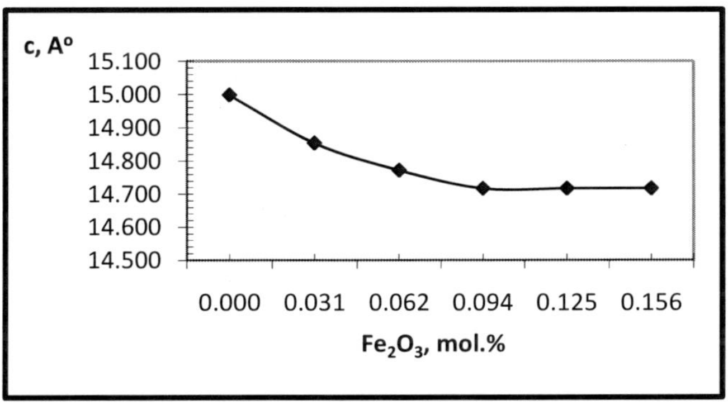

(b)

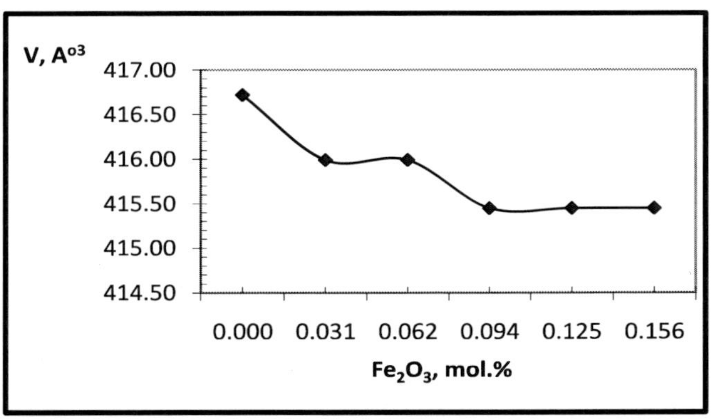

(c)

Figure 17. Unit-cell parameters (a) *a*, (b) *c*, and (c) *V* of Al_2O_3 as functions of iron oxide content.

The oxidation of divalent iron and precipitation of aluminum iron hydroxycarbonates can be represented by the following schemes:

$$2FeSO_4 + (NH_4)_6Mo_7O_{24} + H_2O = Fe_2(SO_4)_2(OH)_2 + (NH_4)_6Mo_7O_{23} \quad (13)$$

$$FeSO_4 + Al_2(SO_4)_3 + 9NH_4HCO_3 = (NH_4)_2Al_2Fe(OH)_5(CO_3)_2 + 4(NH_4)_2SO_4 + 2H_2O + 7CO_2 \quad (14)$$

Based on the theoretical views of olation processes, the tendency of Al^{3+}, Fe^{3+}, and Fe^{2+} to form polynuclear complexes, and the present IR spectroscopy and chemical analysis data, the composition of the ammonium hydroxycarbonates and the oxidation of Fe^{2+} to Fe^{3+} can be represented by the following scheme:

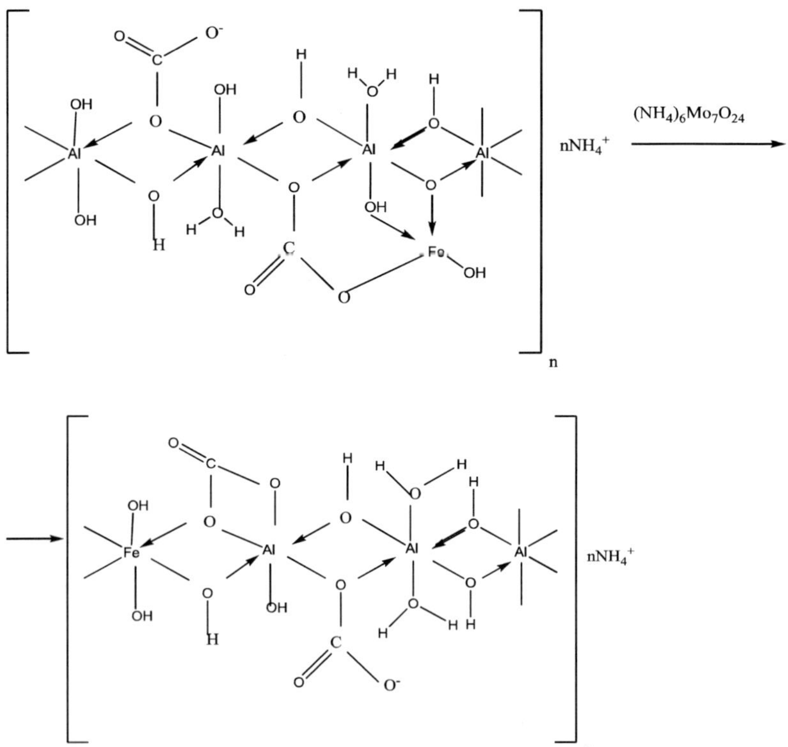

(15)

Heat treatment leads to a decomposition of the ammonium molybdate in the oxidation state 5.7+ (fractional value for summary oxidation number is typical for the Mo- compound [53]) and the formation of tetragonal molybdenum dioxide by the reaction:

$$(NH_4)_6Mo_7O_{23} = 6NH_3 + 7MoO_2 + 3H_2O + 3O_2 \qquad 16)$$

The optimal calcination temperature for achieving high polishing performance is 1150–1200°C. Isothermal holding leads to the formation of two rhombohedral (sp. gr.) solid solutions: α-$Al_{2-x}Fe_xO_3$ (*hkl* 012, 104, 110, and 116 diffraction lines) and $Fe_{2-y}Al_yO_3$ (*hkl* 104, 110, and 116 lines). There is also tetragonal (sp. gr. *P42/mnm*) molybdenum dioxide, MoO (*hkl* 110, 200, and 211 lines), close in lattice parameters to tetragonal manganese dioxide. Figure 15 shows the XRD patterns of an aluminum iron oxide modified with molybdenum dioxide (Figures 15a, 15b) and unmodified material (Figure 15c). In the latter case, in addition to corundum and hematite based rhombohedral solid solutions, there is a cubic aluminum iron oxide of composition $Al_3Fe_5O_{12}$ (*hkl* 222, 332, and 431), whose content is 3 to 12 wt %, depending on iron oxide content (0.125–0.156 mol %). Figure 17 illustrates the effect of ferric oxide content on the percentages of the $Al_{2-x}Fe_xO_3$ and $Al_{2-x}Fe_xO_3$ solid solutions. The presence of cubic aluminum iron oxide improves the polishing performance of the material for the polishing of quenched steels. The thermal transformations of ammonium hydroxycarbonates with ammonium molybdate additions can be represented by the following scheme:

$$NH_4Al_2Fe(OH)_5(CO_3)_2 + (NH_4)_6Mo_7O_{24} \rightarrow NH_4Al_2Fe(OH)_6(CO_3)_2 +$$

220-250°C 425-450°C

$$+ NH_4Al_2Fe(OH)_5(CO_3)_2 + (NH_4)_6Mo_7O_{23} \rightarrow Al_2Fe(OH)_4O_2 \rightarrow$$

950-1150°C

$$\rightarrow Al_2O_3 \cdot Fe_2O_{3\,amorph.} + Fe_2O_{3\,cryst.} \rightarrow Al_{2-x}Fe_xO_{3\,cryst.} + Fe_{2-y}Al_yO_{3\,cryst.} + MoO_{2\,cryst} \qquad (17)$$

The formation of solid solutions from hydroxycarbonate complexes with molybdate ion additions influences the abrasive properties of the material.

The presence of 0.001–0.005 mol % MoO_2 increases the abrasion rate for copper and brass by a factor of 6, as illustrated by the kinetic curves in Figure 8. The highest abrasion rate is offered by the samples containing 0.15–0.18

mol. % Fe_2O_3 at corundum: a hematite molar ratio of 4:1. The lowest surface roughness value, R_a = 0.002 μm, was obtained with the samples containing 0.125–0.156 mol. % ferric oxide.

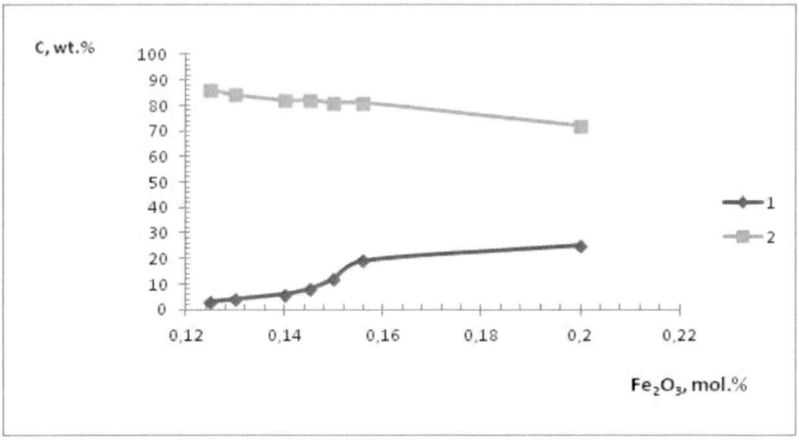

Figure 18. Effect of Fe_2O_3 content on the percentages of (1) hematite- and (2) corundum-based rhombohedral solid solutions.

The minimum surface roughness value was 0.015 μm as determined by optical interferometry. Figure 6 illustrates the polishing effect on the surface topography of copper specimens. According to AFM results, R_a = 0.002–0.005 μm. Particle-size analysis data show that, in the aluminum iron oxide samples modified with molybdenum dioxide, the fraction of particles 10 nm or smaller in size is the same as in the unmodified samples (4–5%). The main particle size fraction is 50 nm.

CONCLUSION

Therefore, we considered mechanochemical regularities of final polishing with the application of abrasive material characterized by tribochemical activity based on solid solutions of oxides of aluminum and iron, as well as modified by molybdenum or manganese, containing nanodispersed particles. It is demonstrated that the process which determines the velocity and quality of metal polishing is the formation of oxide film on the surface and its removal. Equations for heterogeneous processes are used to describe the polishing kinetics. The achieved surface roughness of hardened steel and nonferrous metals R_a is less than 0.005 μm in one finishing procedure.

Aluminum iron oxide nanoparticles were synthesized from ammonium hydroxycarbonate complexes with the general formula $(NH_4)_2Al_2Fe(OH)_5(CO_3)_2nH_2O$ and were characterized by XRD, IR spectroscopy, DTA, electron microscopy, and particle size analysis.

The nanoparticles are shown to consist of α-$Al_{2-x}Fe_xO_3$ corundum-based solid solutions and α-$Fe_{2-y}Al_yO_3$ hematite-based solid solutions. The synthesized mixed oxide offers high polishing performance, as demonstrated in polishing of the ShKh-15 quenched steel and ensures surface roughness values R_a less than 0.005 μm, which is due to its enhanced tribochemical activity.

We studied the influence of molybdenum dioxide on the formation of α-$Al_{2-x}Fe_xO_3$ and α-$Fe_{2-y}Al_yO_3$ solid solutions during heat treatment of ammonium hydroxycarbonate complexes. The polishing performance of the modified materials for final polishing of nonferrous metals and alloys (copper and brass, aluminum) is a factor of 6-7 higher than that of unmodified aluminum iron oxides, which is attributable to the increase in the chemical activity of the abrasive material. The surface roughness below 0.005 μm and the high abrasion rate are due to the oxidizing power of the abrasive particles.

ACKNOWLEDGMENTS

This work was supported by the Russian Program of Basic Research of State Academy of Science from 2008-2012.

REFERENCES

[1] Artemov, A. S. (2004). Polishing Nanodiamonds. *Solid State Phys.*, 46, P. 687.

[2] Jackson, M. J., Mills, B., Hitchiner, M. P. (2003). Controlled Wear of Vitrified Abrasive Materials for Precision Grinding Applications. *Sadhana,* 28, 897-914.

[3] Jackson, M. J., Hitchiner, M. P. (2013). High Performance Grinding and Advanced Cutting Tools. SpringerBriefs in Applied Sciences and

Technology, Chapter 1. Abrasive Tools and Bonding Systems, New York, pp. 1-43, ISSN 2191-530X.

[4] Bakharev, V. P. (2009). Dispersion of Ceramics and Composites in Diamond Finishing by Free Abrasive. Russ. Eng. Res., 29, 162-168.

[5] Filatov, Yu. D. (2009). Assessment of Surface Roughness and Reflectance of Nonmetallic. Products upon Diamond Abrasive Finishing. Sverkhtverd. *Mater.,* No. 5, 70-81.

[6] Storozhenko, P. A., Guseinov, Sh. L., Malashin, S. I. (2009). Nanodispersed Powders: Synthesis Methods and Practical Applications. Nanotechnol. Russia, 4, P. 262.

[7] Pokropivnyi, V. V., Silenko, P. M. (2006). Silicon Carbide Nanotubes and Nanotubular Fibers: Synthesis, Stability, Structure and Classification. *Theor. Exp. Chem.,* 42, 3-15.

[8] Khanra, A. K. (2007). Production of Boron Carbide Powder by Carbothermal Synthesis of Gel Material. *Bull. Mater. Sci.,* 30, 93-96.

[9] Bogdanov, S. P. (2008). Influence of Superstoichiometric Boron on the Synthesis of Cubic Boron Nitride. *Glass Phys. Chem.,* 34, 336-339.

[10] Degtyareva, E. V., Kabakova, I. I., Skorodumova, E. B., Armyanovskii, V. E. (1982). An Abrasive Resistant Corundum Ceramic for Drawing Microwires. Refract. Industr. *Ceram.,* 23, 82-87.

[11] Song, X.-l., Qu, P., Yang, H.-p., He, X., Qiu, G.-z. (2005). Synthesis of γ-Al_2O_3 Nanoparticles by Chemical Precipitation Method. J. Cent. South. Univ. *Technol.,* 12, 536-541.

[12] Kotov, Yu. A., Samatov, O. M. (1994). Characteristics of Aluminum Oxide Powders Produced by Pulsed Heating of a Wire. Poverkhnost., no. 10–14, 90-94.

[13] Safronov, A. P., Kalinina, E. G., Blagodetelev, D. A., Kotov, Yu. A. (2010). Separation of Aluminum Oxide Powders with Different Degrees of Aggregation by Sedimentation in an Aqueous Medium. Nanotechnol. Russia, 5, no7-8, 498-505.

[14] Belousov, V. M., Chertov, V. M., Rozhkova, E. V., Litvin, V. I., Zazhigalov, V. A. (1997). A Sol-Gel Method for Synthesizing Porous Iron Aluminum Oxide Substances and Regulation of their Physicochemical Characteristics. *Teor. Eksp. Khim.,* 33, 120-123.

[15] Silyakov, L., Pesotskaya, N. S., Yukhvid, V. I. (1995). Selfpropagated High-Temperature Synthesis and Properties of Corundum-Based Abrasive Composition Material. *Inorg. Mater.,* 31, 351-357.

[16] Tarasov, A. G., Gorshkov, V. A., Yukhvid, V. I. (2007). Phase Composition and Microstructure of Al_2O_3 - Cr_2O_3 Solid Solutions

Prepared by Self-Propagating High-Temperature Synthesis. *Inorg. Mater.,* 43, P. 724.

[17] Merzhanov, A. G., Borovinskaya, I. P., Prokudina, V. K., Pesotskaya, N. S., Nasonova, M. A. (1998). SHS-Abrasives: Production, Properties, Application. Nauka Proizvodstvu, no. 8, 4-12.

[18] Kimi, N.-H., Lim, J.-H., Kim, S.-Y., Chang, E.-G. (2005). Semi-Abrasive Free Slurry with Acid Colloidal Silica for Copper Chemical Mechanical Planarization. J. Mater. Sci.: *Mater. Electron.,* 16, 629-632.

[19] Viktorov, V. V., Fotiev, A. A., Badich, V. D. (1996). Abrasive and Thermal Properties of Al_2O_3–Cr_2O_3 Solid Solutions, *Inorg. Mater.,* 32, no. 1, 63-65.

[20] Chekhomova, L. F. (1999). Abrasive Performance of Modified Cr_2O_3. *Fiz. Khim. Obrab. Mater.,* no. 5, 88-94 (in Russian).

[21] Chekhomova, L. F. (2001). Abrasive Properties of Modified Chromia. Inorg. *Mater.,* 37, no. 3, 274-280.

[22] Koroleva, L. F. (2002). Modified Chromium Oxides, Oxyhydroxide, and Spinels for Abrasive Materials and Pigments. Ekaterinburg: Rus. Academy Sciences, Branch Ural., 217 p.

[23] Koroleva (Chekhomova), L. F. (2003). Utilization of Chromia Wastewater Sludge in the Technology of a New Abrasive Material. Khim. Interesakh Ustoich. *Razvit.,* 11, no. 3, 509-513.

[24] Maslov, E. N. (1974). Theory of grinding materials. Moscow. Mechanical Engineering, 319 p.

[25] Yashchcritsyn, P. I., Martynov, A. N. (1983). Finishing of Details in Engineering. Minsk, Vysh. Shkola, 191 p.

[26] Rebinder, P. A. (1940). Bulletin of the Academy of Sciences of the USSR, no. 8, 5-28.

[27] Hajnike, G. (1987). Tribochemistry. Moscow. Mir, 584 p.

[28] Grebenshchikov, I. V. (1935). Role of chemistry in polishing processes. Soc. reconstruction and science, 2, 22-33.

[29] Rajan, Krishna (1996). Chemical-Mechanical Polishing of Oxide Thin Films: The Rebinder-Westwood Phenomenon Revisited, *Journal of Electronic Materials,* 25, no. 10, 1581-1584.

[30] Lihtman, I., Schukin, E. D., Rebinder, P. A. (1962). Physico-chemical mechanics of metals. Moscow. Academy of Sciences of USSR, 303 p.

[31] Khodakov, G. S., Kudryavtseva, N. I. (1985). Physical and chemical bases of polishing a glass. Moscow: Nauka, 185 p.

[32] Khodakov, G. S. (2000). Physico-chemical mechanics technology material processing. *Rus. Chem. Jornal.,* no. 3, 93-107.

[33] Koroleva, L. F. (2009). Abrasive Properties of Aluminum Iron Oxide Nanoparticles. *Inorg. Mater.,* 45, no. 10, 1158-1164.

[34] Koroleva, L. F. (2006). Tribochemical Activity of Mixed Oxides Abrasive Materials in a Metal Polishing. *Fiz. Khim. Obrab. Mater.,* no. 4, 84-92.

[35] Rogov, V. V., Rublev, N. D., Krotenko, T. L., Troyan, A. V. (2008). A Study of Intensity of Tribochemical Contact Interaction between a Polishing Compound and Sapphire in Machining. *Sverkhtverd. Mater.,* no. 4, 75-78.

[36] Kostetskii, B. I., Nosovskii, I. G., Karaulov, A. K., et al. (1976). Surface Strength of Materials in Friction. Kiev, Tekhnika, 292 p.

[37] Marchenko, E. A. (1979). On the Nature of Fracture of Metal Surface in Friction. Moscow, Nauka, 118 p.

[38] Migachev, B. A. (2001). Identification of Damage at Metal Deformation. Ekaterinburg, UrO RAN, 195 p. (in Russian).

[39] Lin, Y.-Y., Lo, S.-P. (2004). A Study of a Finite Element Model for the Chemical Mechanical Polishing Process. *Int. J. Adv. Manuf. Technol.,* 23, 644-650.

[40] Wang, Y. G., Zhao, Y. W., Li, X. (2008). Modeling the Effects of Abrasive Size, Surface Oxidizer Concentration and Binding Energy on Chemical Mechanical Polishing at Molecular Scale. *Tribol. Int.,* 41, 202-210.

[41] Bibik, E. E. (2010). Mechanochemistry of Metal Polishing with an Abrasive Suspension. *Russ. J. Appl. Chem.,* 83, 811-815.

[42] Koroleva, L. F. (2010). Synthesis and Abrasive Properties of Nanoparticulate MoO_2 Modified $Al_{2-x}Fe_xO_3$ $_a$nd $Fe_{2-y}Al_yO_3$ Solid Solutions. *Inorg. Mater.,* 46, no. 12, 1330-1336.

[43] Koroleva, L. F. (2012). Final Polishing of Metals to Obtain Nanoroughened Surface. *Nanotechnologies in Russia.,* 7, no. 1-2, 67-75.

[44] Pout, J., Tu, K., Meyer, J. (Eds.) (1982). Thin films. Interdiffusion and reactions. Moscow. Mir, 575 p.

[45] Cabrera, N. (1956). The oxidation of metals. In: Semiconductor surface physics. Univ. Pennsylvania Press, 327-332.

[46] Mott, N. (1949). The theory of protective films formation on the metal surface. *Trans. Faraday Soc.,* v.40, 472-486.

[47] Myagkov, V. G., Bykova, L. E., Li, L. A., et al. (2002). Solid Phase Reactions, Self Propagating High Temperature Synthesis, and Martensitic Transformations in Thin Films. *Dokl. Phys.,* 47, no. 2, 95-98.

[48] Boldyrev, V. V. (1997). Reaction Ability of Solids, Sib. Branch RAN, Novosibirsk, 304 p.
[49] Vinogradova, I. E. (1972). Protivozadirnye prisadki k maslam (Antiscoring Additives to Oil), Moscow: Khimiya, 272 p.
[50] Koroleva (Chekhomova), L. F. (2004). Synthesis of Spinel-Based Ceramic Pigments from Hydroxycarbonates. *Glass and Ceramics,* 61, no. 9-10, 299-302.
[51] Koroleva, L. F. (2012). Synthesis of nano-dispersed functional materials. Pigments. Abrasives. Sorbents. Biomaterial. Palmarium Academic Publishing, 96 p. ISBN 978-3-8473-9244-6, (in Russian).
[52] Nakamoto, K. (1963). Infrared Spectra of Inorganic and Coordination Compounds, London: Wiley, 411 p.
[53] Little, L. H. (1967). Infrared Spectra of Adsorbed Species. New York: Academic, 514 p.
[54] Toropov, N. A., Barzakovskii, V. P., Lapin, V. V., Kurtseva, N. N. (1965). Diagrammy sostoyaniya silikatnykh sistem: Spravochnik (Phase Diagrams of Silicate Systems: A Handbook), issue 1: Dvoinye sistemy (Binary systems). Moscow: Nauka, part 1.
[55] Tsvigunov, A. N., Apolenis, A. V., Annikov, V. E. and Raikova, V. M. (2007). Detonation Synthesis of Iron Aluminates and a New Form of Alumina. *Glass and Ceramics,* no. 12, 15-22.
[56] Busev, A. I. (1962). Analiticheskaya khimiya molibdena (Analytical Chemistry of Molybdenum). Moscow: Akad. Nauk SSSR, 301 p.

In: Recent Advances in Abrasives Research
Editor: Dirk Bähre
ISBN: 978-1-62808-566-2

Chapter 7

ONE-WAY ABRASIVE FLOW MACHINING: BASIC PARAMETER INFLUENCES AND IN-PROCESS AXIAL FORCE MEASUREMENT

Horst Bruennet** *and Martin Swat
Saarland University, Institute of Production Engineering,
Saarbruecken, Germany

ABSTRACT

Abrasive flow machining (AFM) is an abrasive machining process used to improve the surface quality of machined components and to reduce stress concentrations at failure critical areas. The silicate-based carrier medium loaded with abrasive particles is used to machine inaccessible areas and complex internal passages of the corresponding work pieces. Application examples are components of fuel injection systems which are treated with this process in order to withstand higher internal pressures. On the one hand, this chapter delivers insight into results of a recent parameter study in one-way abrasive flow machining. After presenting a brief overview of related research papers, new results regarding the influence of medium pressure and lead time on the form tolerance, surface quality and wear of both the work piece and the adapter are discussed. On the other hand, an in-process axial force measurement setup is presented in order to measure the influence of the applied medium pressure on the machined part of the work piece. The results

* Corresponding author. Email: lft@mx.uni-saarland.de

cover two sample work piece materials: a commonly used automotive steel AISI 4140 and a polyamide PA 6.

Keywords: Abrasive flow machining, finishing, deburring, force measurement

1. Introduction

Hydraulic components are often subjected to pulsating pressure loads. As a consequence, the surface integrity of such components plays a key-role for the fatigue resistance. This is particular true for components with complex internal geometries, e.g. bore intersections. The stress concentrations at such failure critical areas need to be minimized in order to avoid crack initiation. The strength of the surface can be improved by appropriate polishing processes, which effectively reduce the surface roughness. Additionally, the mechanical surface treatment induces compressive residual stresses which are beneficial in terms of the fatigue resistance. KLOOS et al. [1] give an interesting literature overview of the influence of surface polishing on fatigue strength. Abrasive Flow Machining (AFM) is a high performance polishing process capable of achieving very good and reproducible surface finishes. At the same time, stress concentrations at bore intersections can be reduced by generating defined edge rounding. This is used to optimize the hydraulic flow during the lifetime of the treated components in order to avoid cavitation effects which would otherwise lead to severe damage. Popular application examples are components of common rail diesel injection systems. They have to withstand pressures of 2,200 bar and higher for at least 250,000 km.

The first part of this chapter gives a brief overview of related research papers. Afterwards, results from recent parameter studies in one-way AFM are presented and discussed. Considered parameters are medium pressure and lead time for a given medium with aluminum oxide abrasives. Their influence on form tolerance such as roundness or straightness, surface quality and wear of both the work piece and the adapter are investigated. For the first time, long time machining effects are discussed and suggestions for future process control are given at the end of the parameter study. The results cover two sample work piece materials: a commonly used high strength automotive steel AISI 4140 and a polyamide PA 6. Then, a novel in-process axial force measurement setup is presented in order to measure the influence of the applied medium pressure on the machined part of the work piece. The complex and hard to predict frictional behavior of the medium during machining is empirically described with the help of the axial force measurements. Future numerical calculations can use this information to improve the material description used to predict the behavior during machining.

2. State of the Art

AFM is an abrasive cutting process where a polymer-based pliable carrier medium laden with abrasive particles is forced through hard-to-reach internal geometries. The medium is often referred to as "liquid sandpaper" or "honing stone". Both metaphors properly describe the viscoelastic behavior of the medium. On the one hand, the medium tends to flow when it is slowly loaded with pressure. On the other hand, it is rigid like a stone when it is quickly loaded with pressure. Even though the actual cutting is done by the abrasives, the property of the carrier medium is crucial as it causes the abrasives to cut [2]. AFM can be applied with one-way or two-way medium flow. In one-way systems the medium is extruded through the work piece and exits the work piece after each cycle. Cutting forces work in only one direction. In contrast, in two-way systems two vertically opposed cylinders flow the medium back and forth through the work piece. Depending on the respective flow system, a cycle is usually defined as an extrusion of the medium volume that is done once or twice. For two-way AFM the number of cycles is cited to range from 1 to 100 [3]. The main process parameters of AFM can be classified as parameters of the work piece, the medium and the parameters of the machine [4].

Work Piece Parameters

Parameter studies have been carried out for a number of materials ranging from soft metals like aluminum and brass [5-7] and various steel grades [8-13] to ceramic materials [14]. The results are discussed in the sections *medium parameters* and *process parameters* below.

Medium Parameters

The main parameters of the AFM medium are the viscosity of the carrier medium and the type, size and concentration of the abrasives. The viscosity of the carrier medium should be chosen depending on the application. RHOADES [15] proposed the application of low viscous media to achieve selectively higher material removal for deburring and rounding of edges and intersections. Uniform material removal can be achieved with higher viscous media. A higher viscosity of the medium also leads to higher removal rates and better surface finish [16]. Frequently used abrasives in AFM media are aluminum oxide, boron and silicon carbide or diamond which influence the removal rate and surface finish [2, 6, 17]. Regarding the size of the abrasives, parameter studies show that the material removal rate can be increased with increasing size of the abrasives [5]. However, the size of the abrasive particles also limits the achievable surface finish. Large particles compared to the initial surface roughness can even deteriorate the surface finish [18]. Better surface finish can be achieved with smaller abrasives [19]. Higher concentration of abrasives can also lead to better surface finish for a certain number of cycles [10, 18].

Process Parameters

The adjustable piston pressure and the number of performed cycles are the most important machine parameters. Industrial AFM machines are available with piston pressures adjustable from 7 to 220 bar [6]. In literature most investigations document piston pressures in a range of 5 to 80 bar [6-9, 19-21]. The processing time is always dependent of the number and definition of a cycle, i.e. one-way or two-way cycles and medium volume per cycle. Both parameters have a strong influence on the performance and results of the AFM process like the material removal, the surface improvement per AFM cycle and the achievable surface finish.

JAIN et al. varied the piston pressure between 3 and 14 bar. For a fixed number of cycles they found that with higher piston pressure the surface roughness can be improved to a certain critical surface roughness. They state that with further increase of the piston pressure the surface finish cannot be further improved or it can even worsen due to the increased cutting depth of the abrasive particles [18].

RAJU et al. use a laboratory AFM set-up with a piston pressure of 10 bar [21]. They also find a strong improvement of surface roughness within the first few AFM cycles, followed by a moderate improvement. Finally, the surface finish deteriorates with an increasing number of cycles. In further investigations by BAEHRE et al. the piston pressure has been varied between 40 and 70 bar in steps of 10 bar and the surface roughness has been documented for an increasing number of cycles [13]. The investigations also show better surface roughness for higher piston pressures for each number of cycles. However, the improvement of surface roughness decreases with an increasing number of AFM cycles and the surface finish seems to strive for a critical value of 1 μm independent of the applied piston pressure and the initial surface roughness. A deterioration of the surface roughness values with an increasing number of cycles was not found.

Deburring and Rounding

AFM is not only applied for the smoothing of the surface but also for deburring and rounding difficult to access and failure critical edges and intersections. Burrs are accurately ground away rather than smeared off or broken off [2]. A model for the removal of burrs is given by SZULCZYNSKI [9]. The model describes three mechanisms that appear in deburring with two-way AFM. During the first piston lift, the burr offers resistance and the medium bends the burr in flow direction at the thinnest point. The burr finally breaks off due to back-bending after reversal of the flow direction or due to increased material removal at the root of the burr. The tertiary mechanism is characterized by continuous material removal and edge rounding at the burr root. PRZYKLENK found that the rounding will not result in perfectly shaped quarter-circle radii that tangentially merge into the bordering areas. Instead, a tear-off edge followed by a washout can be the result of the entering medium. He also stated that the radius yields a critical value after a few cycles that is not significantly changed with additional AFM cycles [3].

Form Deviations

BORCHERS found uneven removal of material, out of roundness and bell-mouthing due to inappropriate fixture design [19]. PRZYKLENK stated that the roundness error of bores is not significantly affected as long as the bore and the fixture are axially aligned. He also documented a conical form of long bores after application of AFM and assumed the pressure decrease along the passage length to be the reason [3]. In contrast, RAJU et al. found an improvement in the out of roundness error [21]. Due to these contrary findings, the influence of AFM on roundness errors will be regarded in this chapter.

Force Measurement and Process Models

In literature, different attempts have been undertaken to describe the AFM process. Measurement and prediction of forces at work during the process play an important role in most models. JAIN et al. developed a finite element (FE) model for the flow of the medium during AFM in order to estimate the stresses and forces at work during the process [18]. The results of the FE model are compared to experimental results of JAIN AND ADSUL [5] who investigated the AFM process for brass and aluminum. The comparison of theoretical and empirical values shows decent results for material removal as a function of piston velocity and percentage concentration. The same applies for a comparison of theoretical and empirical results of the change in surface finish. Underlying assumptions, among others, are that every abrasive particle reaches the same indentation depending on the applied piston pressure. It is also assumed that the load on each particle is constant and equal to the average load.

GORANA et al. developed a theoretical model for the forces per active grain during the abrasive flow machining to investigate the mechanisms of the AFM process. In order to evaluate the number of actively cutting grains, a split cylindrical fixture was loaded with the AFM medium at different piston pressures. The fixture was then split and the medium was investigated with a stereomicroscope for visible and shining grains at the surface of the medium which were defined to be active grains. For the in-process measurement they used a laboratory set-up dynamometer to determine the axial and radial forces [20]. Subsequently, GORANA et al. extended their previous investigations to a model for the mechanics of grain-work piece interaction during AFM that strongly depends on the cutting forces per active grain [22]. Force measurements were carried out in experiments with only the carrier medium and experiments using carrier medium with abrasives mixed into it. The difference in measured forces was attributed to the counted number of active grains. As a result, the forces per active grain and the indentation were derived. The measured values were then compared to theoretically determined values necessary for chip formation. However, the measured indentation and force per grain in the experiments were too small compared to the theoretical values. Hence, GORANA et al. concluded that no chip formation seems to occur according to their results. The dominating mechanisms were therefore elastic deformation associated with rubbing and plastic deformation from ploughing.

In contrast, SZULCZYNSKI reported from his scratching experiments that the abrasives were continuously cutting the material like bonded abrasive grains. In his results rubbing or rolling only occurred under certain conditions during the start phase of the process [9]. Further, he proposed a model for the calculation of the edge rounding. In this context he found a significant influence of piston pressure on the edge rounding. He also pointed out the working temperature of the medium to be an important parameter as it influences the viscosity of the medium and hence the velocity and the cutting forces and indentation. Based on the results of SZULCZYNSKI, UHLMANN et al. proposed to develop a numerical model for the flow behavior of the viscoelastic AFM medium. The numerical model should then allow for setting up a process model that is capable of anticipating work results like surface finish and edge rounding [12].

In SCHMITT [23] a numerical model for the non-Newtonian behavior of the AFM medium was presented. The parameters of the medium were identified with the help of rheometer experiments and the model was validated by means of an independent experiment. Also a simple abrasion model for edge rounding was presented that still has to be proven with empirical data.

3. AFM Experimental Setup and Medium

The investigations were performed by using an ExtrudeHone Vector AFM machine equipped with two vertically-opposed media cylinders. In principle, this machine can be run in one-way and two-way mode. Here, only the lower cylinder was used for the considered one-way AFM.

Figure 1 shows the machine in a cross-sectional view taken from HULL [24] along with the corresponding description and a picture of the real machine used for the experiments. The diameter of the medium cylinder was 100 mm. The medium was extruded to a collecting tray and refilled manually into the lower medium cylinder. Piston pressure was adjustable between 28 and 105 bar.

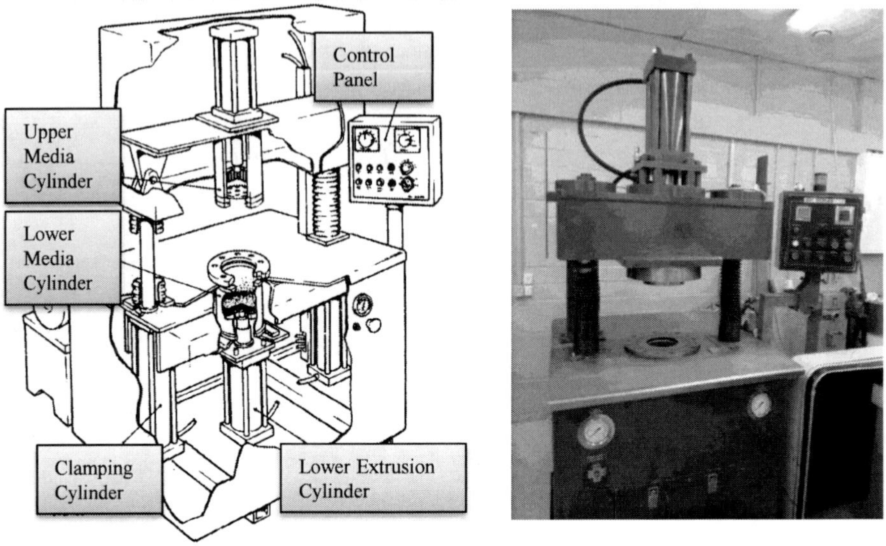

Figure 1. Schematic of an AFM machine (left) taken from HULL [24] and a picture of the used ExtrudeHone Vector AFM machine (right).

Throughout the entire experimental investigation a ready-made high viscosity medium was applied. This medium is the primary cutting tool in AFM. The available characteristics of the medium and a picture of the abrasive grains are summarized in Figure 2. However, detailed information about rheological properties and used additives were not available by the vendor.

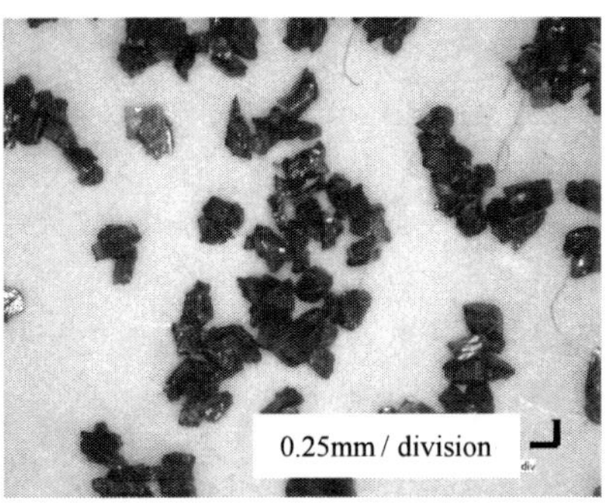

Medium trade name	EM24640
Density	1,9 g/cm3
Material of abrasive grains	Aluminum oxide
Average size of abrasive grains	300 – 600 µm

Figure 2. Available properties and size distribution of abrasive grains.

3.1. Parameter Study in One-Way AFM

The first part of the presented research results contains a parameter study in one-way AFM. Considered parameters are piston pressure and cycle time for the described medium. Their influence on surface quality, form tolerance such as roundness or straightness and wear of both the work piece and the adapter are investigated. The one-way AFM approach has the advantage that process parameters such as cycle time can be documented and its effect on form or surface finish can be analyzed for each AFM cycle. The results cover two sample work piece materials: a commonly used high strength automotive steel AISI 4140 and a polyamide plastic material PA 6. First of all, the used fixture is presented.

3.1.1. AFM Fixture

AFM fixture or tooling is important to clamp the work piece and to guide the abrasive medium. In order to design a flexible fixture, the setup in Figure 3 was used to adapt a variety of different work pieces.

Center pins are used to mount the whole fixture above the lower medium cylinder. The diameter of the main bore in the fixture block is 100 mm and equal to the diameter of the lower medium cylinder. The fixture block is closed with a removable cover plate and clamped between the hydraulic clamp cylinders. Re-filling of the lower medium cylinder after each extrusion cycle is possible by removing the cover plate. In order to ensure a certain flexibility, the outlet of the fixture can be adjusted by using different adapter pieces. A clamping is used to hold each work piece and adapter combination. Then it is fixed to the clamping block by a screw fitting. The medium exits unguided at the outlet side of the part, hence an influence on the roundness value is expected due to bell-mouthing as described in [19].

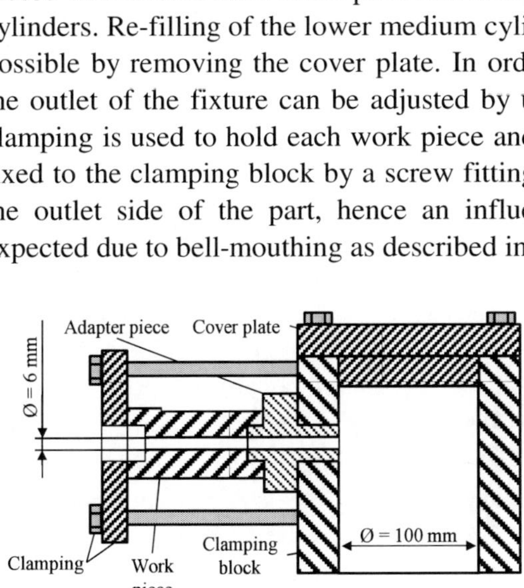

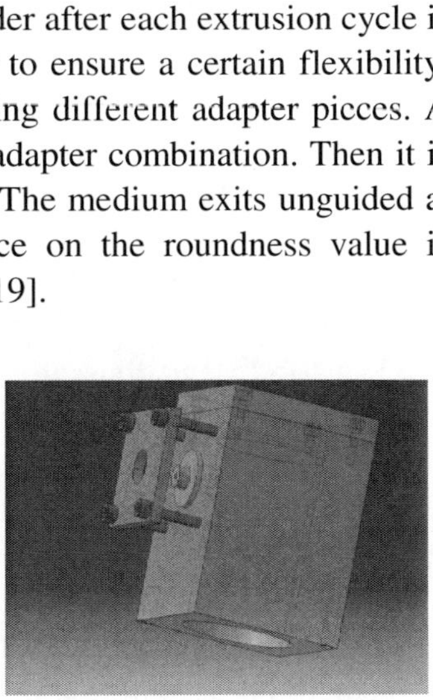

Figure 3. Flexible clamping block for one-way AFM.

3.1.2. Work Pieces

The basic experiments were conducted by using cylindrical work pieces of the high strength steel AISI 4140 with an ultimate strength of 1,073 MPa and Young's modulus of 211,000 MPa. Additionally, polyamide PA6 was used in order to evaluate the wear behavior. PRZYKLENK [3] reported a potential in wear resistance for pressure levels between 14 bar and 30 bar. The geometry of the work pieces included an axial bore of 6 mm in diameter and a passage length (L_P) of 60 mm. The geometry and the medium extrusion during one-way AFM are shown in Figure 4.

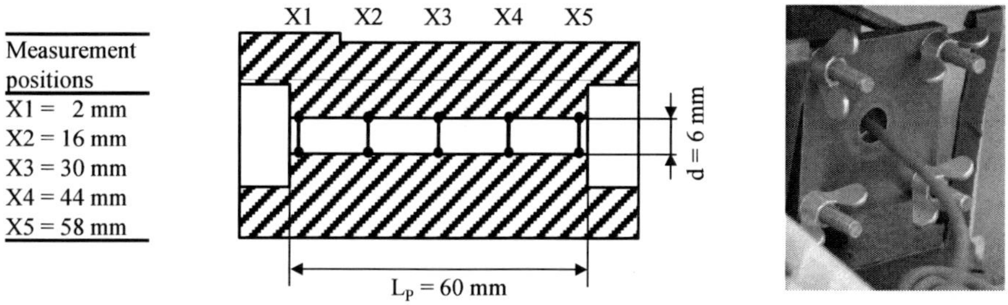

Figure 4. Work piece geometry and medium extrusion during AFM.

3.1.3. Process Chain for Evaluation

In general, an engineering surface and the corresponding evaluation can be divided in a range of spatial frequencies. According to RAJA et al. [25] high frequencies or short wavelength components are referred to as roughness, the medium frequencies as waviness and low frequencies as form. The original surface is a superposition of the first four orders of surface irregularities according to DIN EN ISO 4287 [26] or the comparable ASME standard B46.1 [27].

The form deviations are acquired with the help of a ZEISS metrology system, whereas waviness and roughness are acquired by a Mahr MarSurf surface metrology device. Here, the unfiltered primary profile (P-profile) is separated into the roughness profile (R-profile) and waviness profile (W-profile) by a digital Gauss-filter. The roughness profile is the deviation of the primary profile from the waviness, i.e. R = P-W. This is done automatically by the MarSurf software. As separation boundary serves the so-called cut-off wavelength (λ_c). Here, it was set to 0.8 mm. For the evaluation, only Ra and Rz were considered.

Surface irregularity	Kind of deviation
1st Order: Form 	Straightness, Roundness *Evaluated here: both with scans along the diameter or the length of the passage, respectively*
2nd Order: Waviness 	Waves *Acquired primary profile without roughness*
3rd Order: Roughness 	Grooves and scratches *Evaluated here: Ra, Rz*
4th Order: Roughness 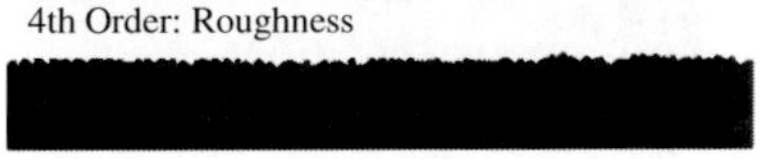	Scores and scales *Evaluated here: Ra, Rz*

Figure 5. Form, waviness and roughness of a surface according to DIN EN ISO 4287 [26].

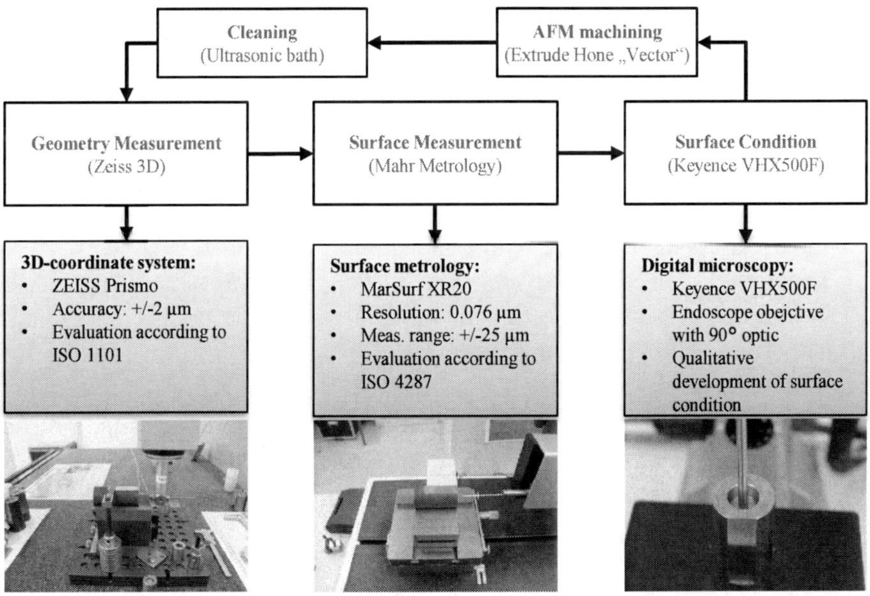

Figure 6. Process chain for evaluation.

The work pieces were machined with a one-way AFM process. After each machining cycle the work pieces were cleaned and measured. In order to clean the work pieces from residuals of the AFM medium, they were put for 10 minutes at 30°C into an ultrasonic bath. The cleaning additive Tickopur R33 was added to achieve a clean surface and to reduce corrosion. Subsequently, the surface roughness of the work pieces was measured using a Mahr MarSurf XR20 surface metrology system. The resolution was as small as 0.076 μm at a measurement range of +/- 25 μm. According to DIN EN ISO 4287, both parameters, Ra and Rz, were measured at five axial measurement positions and at a rotation of 0 or 180 degree, respectively. Then, the diameter was measured with circular scans at five measurement positions as displayed in Figure 4. Here, a ZEISS Prismo 3D-metrology machine with an accuracy of +/- 2 μm was used. The roundness and straightness was evaluated according to ISO 1101. Finally, the qualitative surface condition was evaluated using a Keyence digital microscope VHX500F with a 90° endoscope optic. The process chain is shown in Figure 6.

3.1.4. Results and Discussion AISI 4140

Results: Form

Eight work pieces were used for the basic investigation. They were subjected to four different AFM piston pressure levels of 40 bar, 50 bar, 60 bar and 70 bar, i.e. two work pieces per pressure level. The work pieces were treated with 42 AFM cycles, with every cycle being equivalent to one full extrusion of the lower medium cylinder or 5 kg of medium, respectively. The cycle time varied significantly with the piston pressure. One cycle with 40 bar took about 100 s whereas a cycle with 70 bar took only about 20 s. Every AFM cycle removes some portion of material inside the bore. Conventional cutting processes often use the removed chip volume per time [cm³/min.]. Therefore, $V_{CHIP,AFM}$ was defined as described in Equation 1 in order to achieve a comparable measure. The widening, as the difference of the passage diameter before and after one cycle AFM machining, is described as $\Delta d_{CYLINDER}$, whereas the machined passage length is defined as L_P. $V_{CHIP,\,AFM}$ is shown in Figure 7 depending on the number of cycles.

$$V_{CHIP,AFM} = \frac{\pi \cdot (\Delta d_{CYLINDER})^2}{4} \cdot L_P \qquad (1)$$

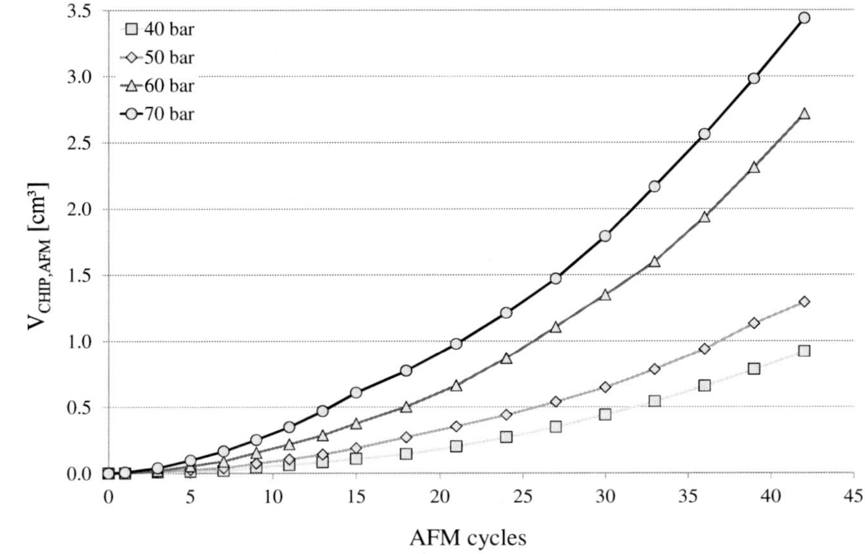

Figure 7. Removed material volume per time.

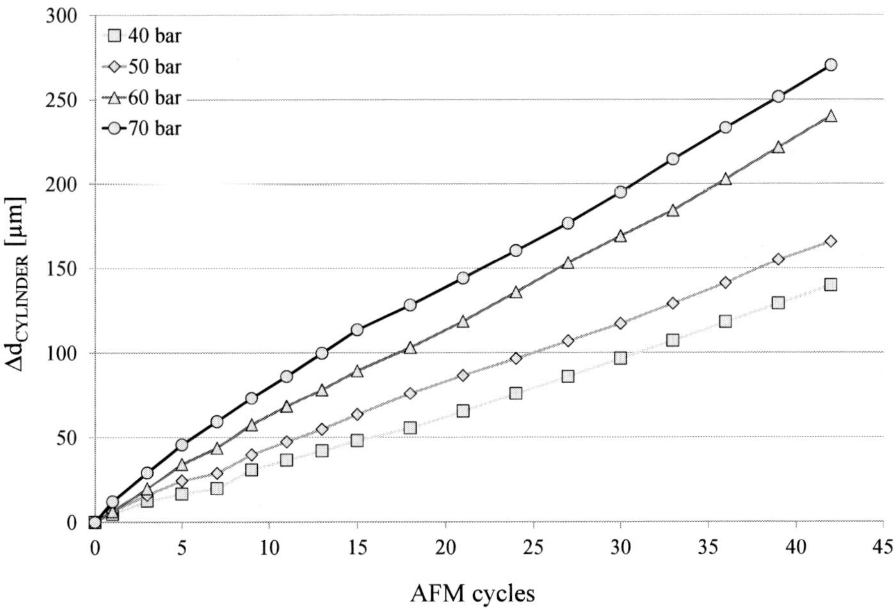

Figure 8. Increase in diameter during AFM.

Chip volume increases as a quadratic function with the number of cycles and the piston pressure. Following Equation 1, this leads to the consequence that the diameter increases linearly with the number of cycles and piston pressure. This behavior can be observed at all measurement positions. Figure 8 shows this behavior.

When applying a linear regression to each of the curves, the gradient gives the widening per AFM cycle. Furthermore, the certainty contains the information about how good the linear regression will fit the curve. Table 1 shows the gradients for each pressure level and the certainty of the regression lines.

Table 1. Gradients and certainty of the linear regression lines for the evaluated widening.

Pressure level [bar]	Gradient [µm/cycle]	Certainty
40	3.26	0.998
50	3.85	0.999
60	5.56	0.999
70	6.16	0.996

The presented gradients indicate the continuous wear of the investigated steel type for the given initial diameter of 6 mm. Hence, those values can be used to predict the lifetime of adapters and tooling which are exposed to the AFM medium and have a comparable geometry.

As the medium flows along L_P, it is assumed that the material removal over the diameter is uniform, i.e. the out-of-roundness error should not be affected. Contrary findings have been presented in literature, therefore measurement positions X1 to X5 were investigated. Figure 9 presents a 3D-plot of the out-of-roundness error depending on measurement position and AFM piston pressure. The out-of-roundness error is calculated as difference between the measured roundness before AFM and after 42 cycles of AFM. Consequently, an increase of this value indicates a worsening, whereas a decrease indicates an improvement. Two work pieces were summarized per pressure level.

It can be observed that the out-of-roundness error worsens along the passage length from inlet to outlet side (X1 to X5). Due to the unguided medium draining at the end of the passage, a so-called bell-mouthing effect is observed in accordance to [19]. Additionally, the out-of-roundness error worsens with increasing piston pressure. Overall, the influence of measurement position and piston pressure is small but existent. When roundness values become important for the function of the passage, special care should be taken in fixture design and lower piston pressures should be applied.

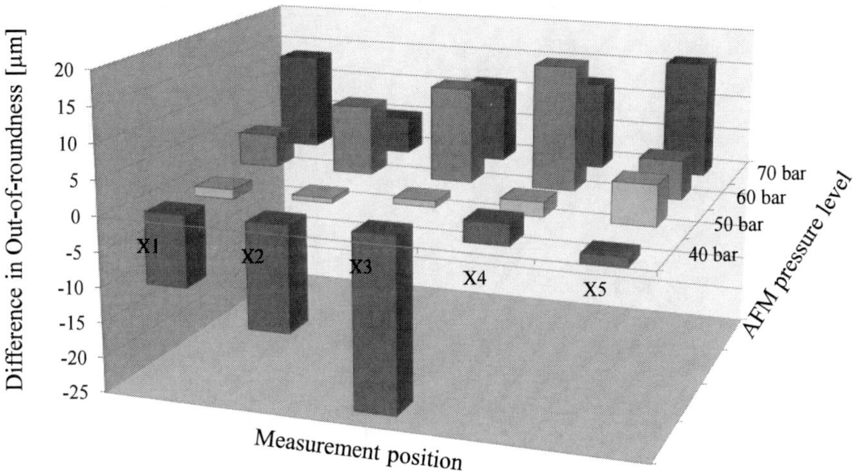

Figure 9. Out-of-roundness error depending on measurement position and piston pressure.

Another aspect of the form evaluation is the straightness of the passage. In order to analyze the behavior during AFM, four line scans were performed over the diameter, i.e. one line scan every 90°. Due to the limited length of the scanning tip, the passage was divided in two work pieces in order to allow the scanning tip to measure from two sides. Hence, the out-of-straightness value is a combined measure and calculated as difference between the initial value before AFM and the final value after AFM. Consequently, an increase of this value indicates a worsening whereas a decrease indicates an improvement. Two work pieces were summarized per pressure level.

Figure 10 presents a 3D-plot of the out-of-straightness error depending on measurement position and AFM piston pressure.

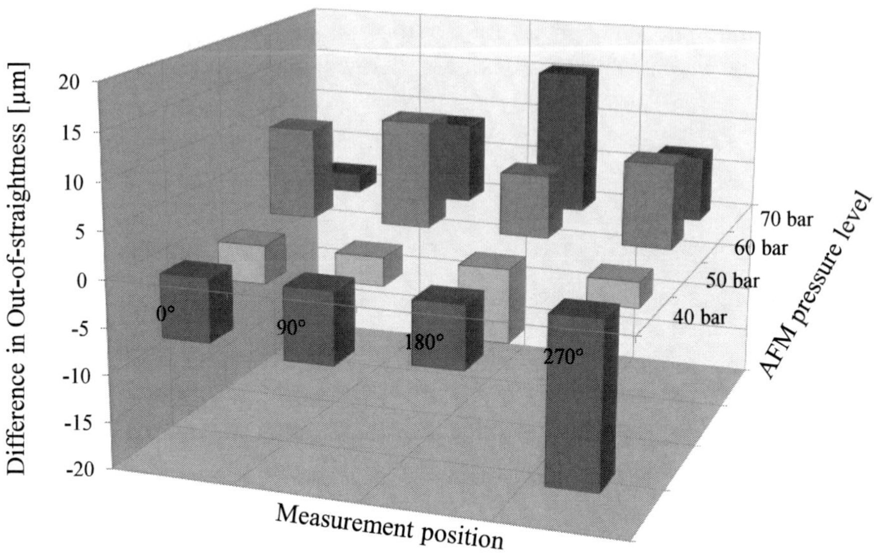

Figure 10. Out-of-straightness error depending on measurement position and piston pressure.

There seems to be an improvement for the out-of-straightness error at smaller piston pressures along with a worsening for higher piston pressures. The position over the diameter does not significantly affect the value.

Overall, the form is affected with higher piston pressures as could be shown for the roundness and the straightness. Additionally, higher piston pressures will remove more material in the same time. This directly affects the lifetime of wear components such as adapters significantly.

Results: Roughness

Figure 11 shows the influence of the number of cycles on the mean Ra-value for two work pieces at the same pressure level. In order to calculate the mean value, measurements were performed for every part and AFM cycle at measurement positions X1 to X5 and 0° and 180° inside the bore. Therefore, every part was investigated with 10 measurements and every pressure level with 20 measurements, as two work pieces were used per pressure level.

Figure 12 shows the single plots for every part with the corresponding standard deviation. For this diagram, the standard deviation was added to the mean value as positive and negative deviation to define the corresponding error bars for every measurement point.

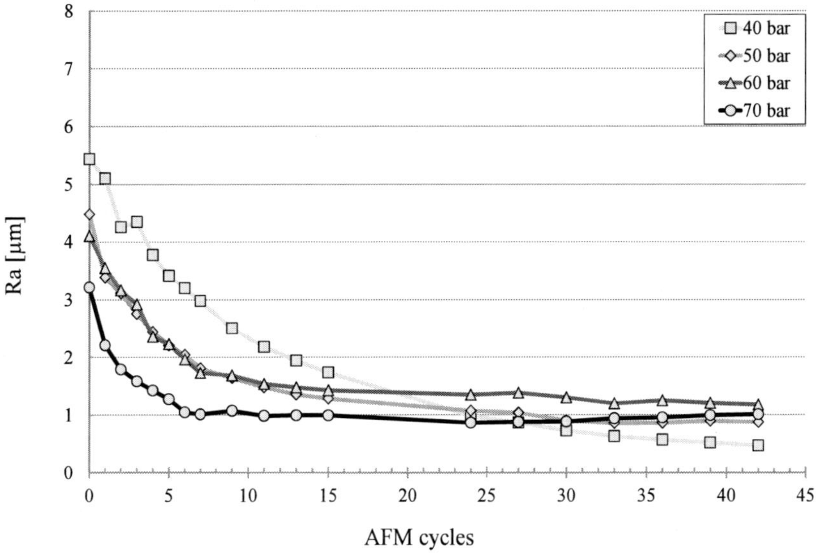

Figure 11. Mean-Ra vs. number of AFM cycles and piston pressure.

In order to identify the influence of the pressure level itself, Figure 13 shows the reduction in Ra by looking at the difference between initial and resulting Ra-values.

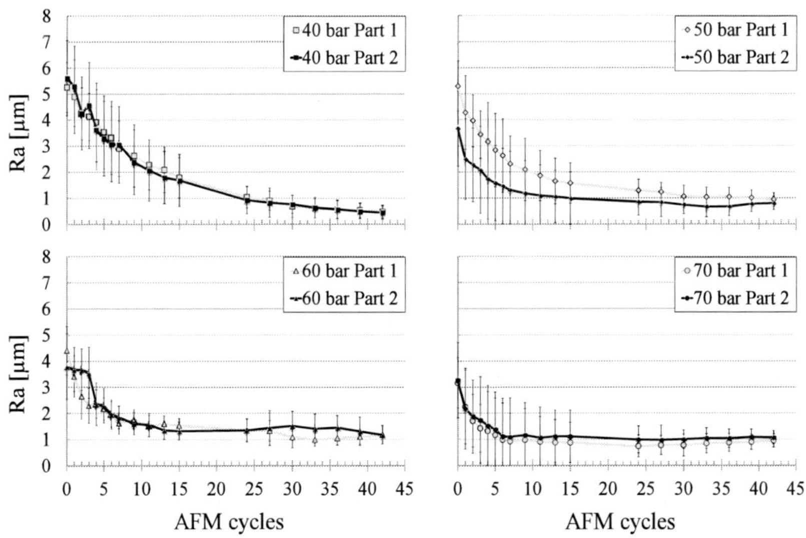

Figure 12. Ra vs. number of AFM cycles and piston pressure for every part.

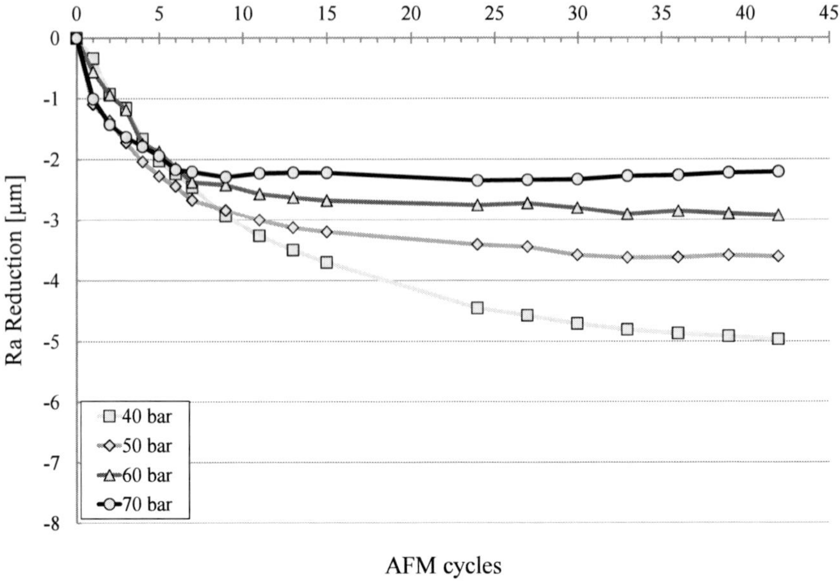

Figure 13. Relative Ra-reduction vs. number of AFM cycles.

AFM with the highest pressure at 70 bar will reach its minimum Ra-value much faster than the other three pressure levels. After 7 cycles, only minor potential in Ra-decrease can be observed. The absolute minimum is reached at around 24 cycles with Ra = 0.86 µm. Interestingly, after reaching this absolute minimum, a further machining will lead to a worsening again. The value after 42 cycles has been steadily increased to 1.1 µm. It becomes also evident that using lower pressure levels shows the potential to reach smaller Ra-values, even though the machining time is much longer. AFM with 40 bar has obviously not reached an absolute minimum yet, even though it has already started to converge. The mean value after 42 cycles is as low as 0.47 µm. From the perspective of cycle time higher pressures lead to much faster results. The same results were obtained for the investigated Rz-values as shown in Figure 14 to Figure 16. The absolute minimum at 24 cycles and 70 bar shows a Rz = 3.57 µm and an increase afterwards. On the other hand, the 40 bar machining reaches a minimum Rz = 1.56 µm. It has to be noted that the initial Ra and Rz values of the work pieces treated with 40 bar were higher than the ones treated with 70 bar. This emphasizes the observation that smaller AFM pressure will potentially yield better surface finishes.

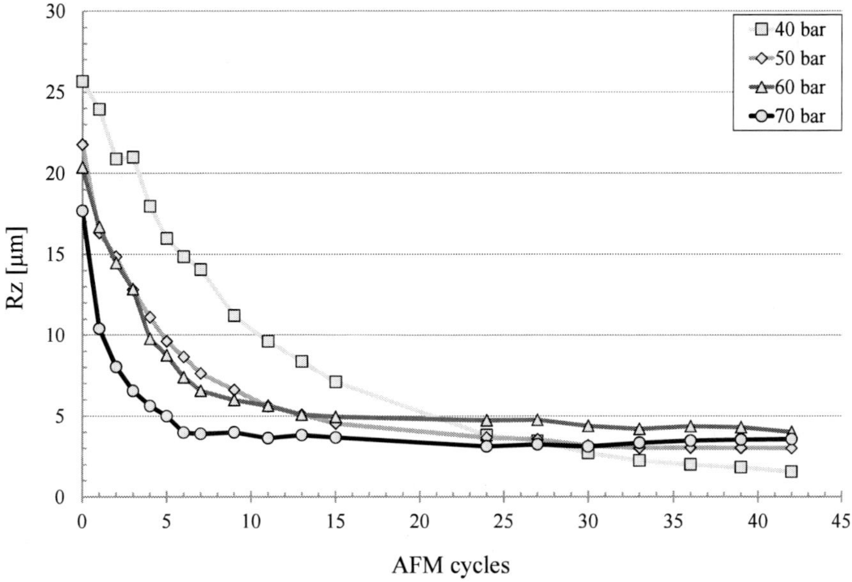

Figure 14. Mean-Rz vs. number of AFM cycles and piston pressure.

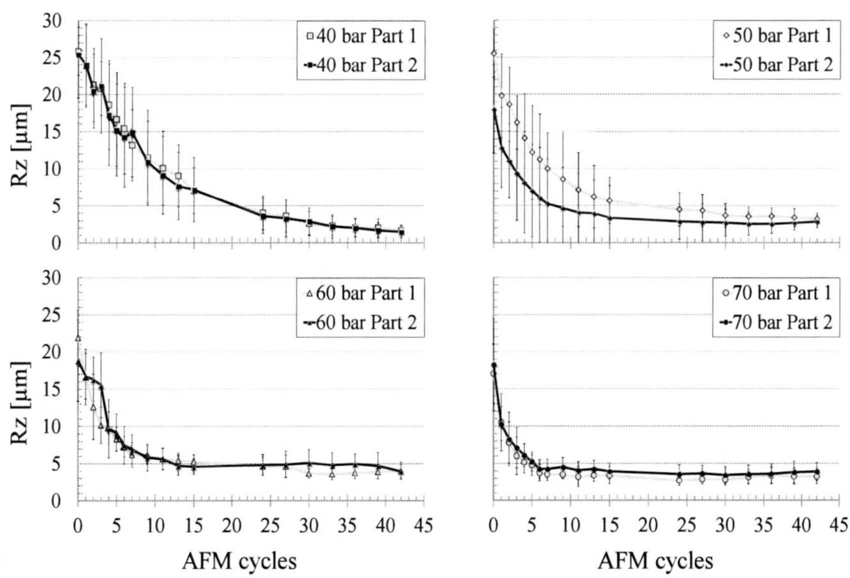

Figure 15. Rz vs. number of AFM cycles and piston pressure for every part.

Figure 12 and Figure 15 show a decreasing standard deviation with increasing AFM cycles and also with increasing piston pressure, i.e. the AFM process leads to a uniform and consecutive smoothing of the surface at all measurement positions. In order to prove this observation, it's worth looking at the single roughness profile and the visual impression of the surface. Figure 17 shows the roughness profile and corresponding pictures made with the digital microscope over the whole machining sequence at five distinct numbers of AFM cycles.

Obviously, the surface is smoothed consecutively over machining time or AFM cycles, respectively. It is also nice to see that the 70 bar pressure level will reach a final surface finish much faster than the 40 bar pressure level – even visually.

This priory acquired information can now be used to suggest an optimization in process control. Given that the formerly found observations are generally true, the AFM process could be optimized by combining different pressure levels in one machining procedure. This can be compared with roughing and finishing in conventional cutting processes. PRZYKLENK [3] showed this behavior for pressure levels between 14 bar and 30 bar and

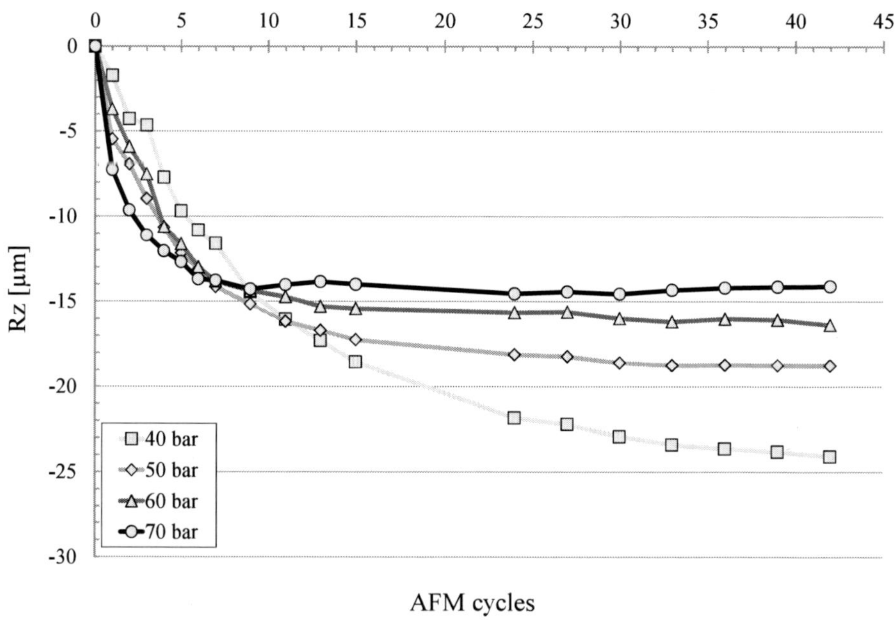

Figure 16. Relative Rz-reduction vs. number of AFM cycles.

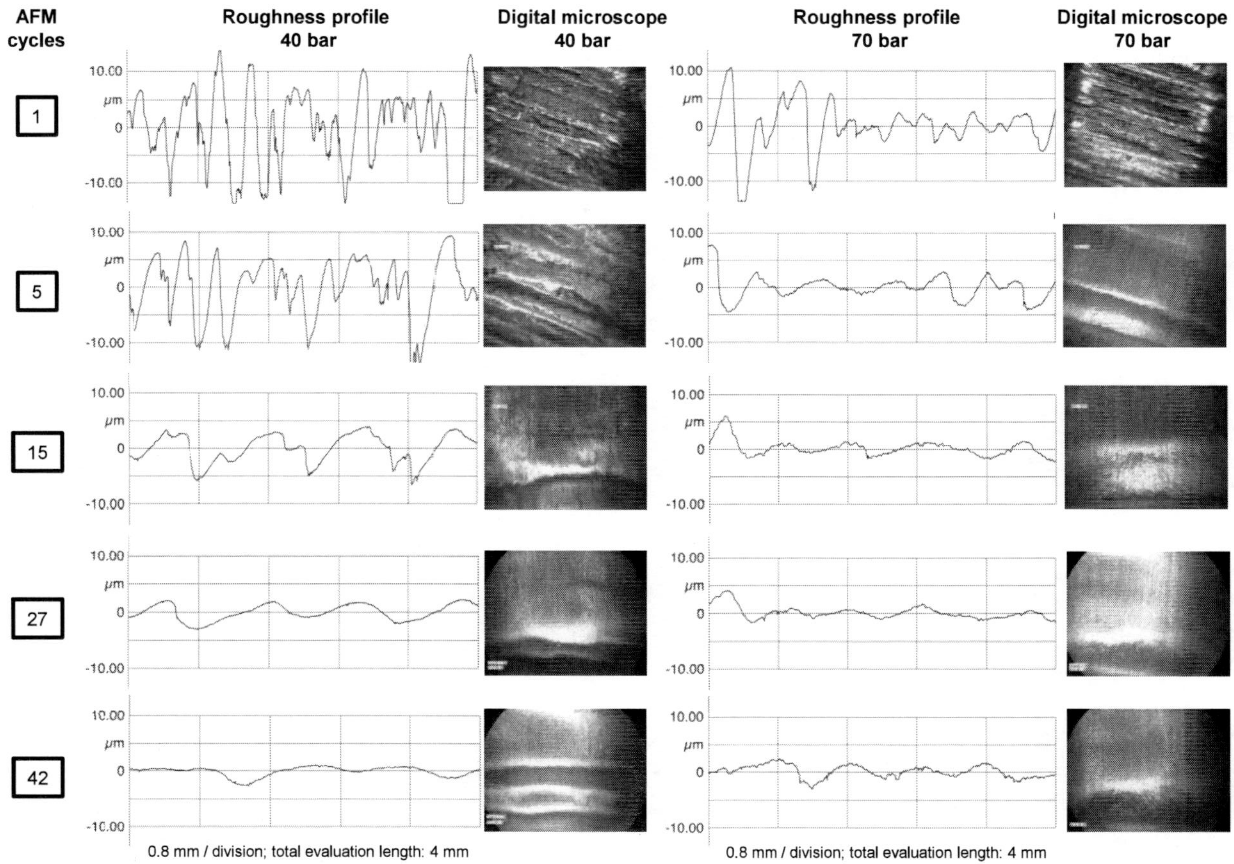

Figure 17. Development of the roughness profile and surface texture with increasing AFM cycles.

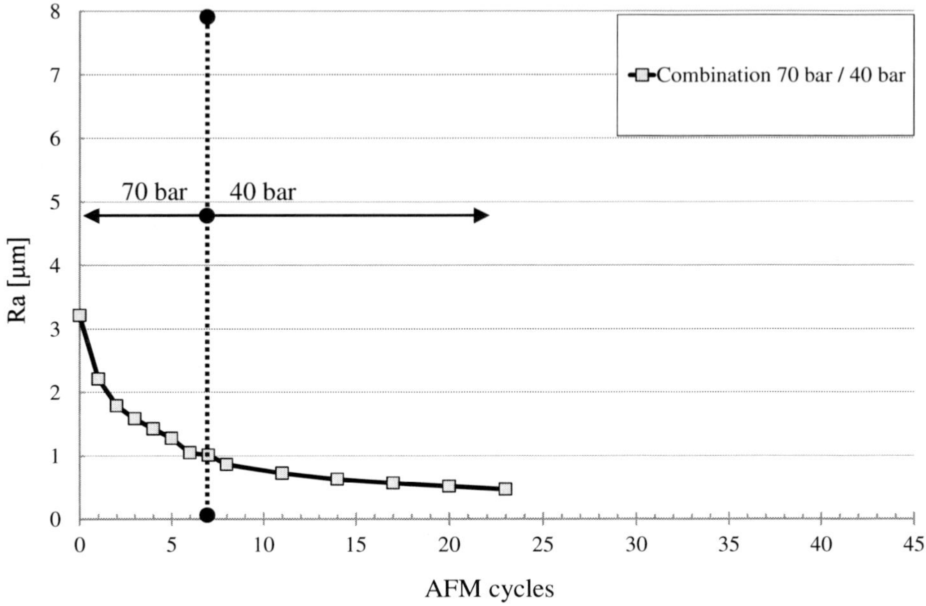

Figure 18. Process optimization by combining different AFM pressure levels.

came to the same conclusion. Taking this into account, the high pressure will be used for a small number of cycles in order to quickly smooth the surface. Afterwards, the pressure is reduced in order to reach the minimum possible surface finish. The idea is sketched in Figure 18. The authors want to emphasize that this idea has to be verified with further experimental data.

When combining the priory acquired Ra-values for the 40 bar and 70 bar pressure levels, the number of AFM cycles could be theoretically reduced by 45 %. More significantly, the machining time could be theoretically reduced by as much as 41 minutes or 58 % when compared to the solely machining at 40 bar. Independently of the actual amount in time saving, the potential of combining different pressure levels is obvious and should be exploited.

3.1.5. Results and Discussion PA 6

Results: Form

AFM is highly demanding in terms of tool wear. Adapter work pieces that are subjected to the AFM medium will deteriorate quite fast, as the results in the preceding investigation for the steel components revealed. The question arises, if a cheaper and easier to machine plastic material would be better suited for the construction of wear components. For this purpose, eight work pieces made of PA 6 were investigated. In this case, only one AFM piston pressure level at 60 bar was used. The work pieces were treated with 22 AFM cycles, as the investigation with the steel components showed a converging behavior for a higher number of AFM cycles. Again, $V_{CHIP, AFM}$ and $\Delta d_{CYLINDER}$ were calculated in order to achieve a comparable measure to conventional cutting processes. The corresponding graphs are shown in Figure 19 and Figure 20. All results are compared to the steel components treated with 60 bar AFM piston pressure.

$V_{CHIP, AFM}$ and $\Delta d_{CYLINDER}$ are comparable for PA 6 and AISI 4140. The plastic material is subject to higher material removal gradients within the first 12 cycles but approximates the behavior of steel afterwards. It can be concluded that the wear behavior of PA 6 is comparable to that of AISI 4140. However, it has advantages over the steel in terms of purchase price, machinability and corrosion resistance. Therefore, it is recommended for the construction of wear components for clamping and adapting purposes. Additionally, the graphs seem to indicate a break-even point after 18 AFM cycles, i.e. the gradient of PA 6 shows a tendency to become smaller than the gradient of AISI 4140. This has to be investigated in future long-term observations as it would support the suggestion to use PA 6 as material for wear components.

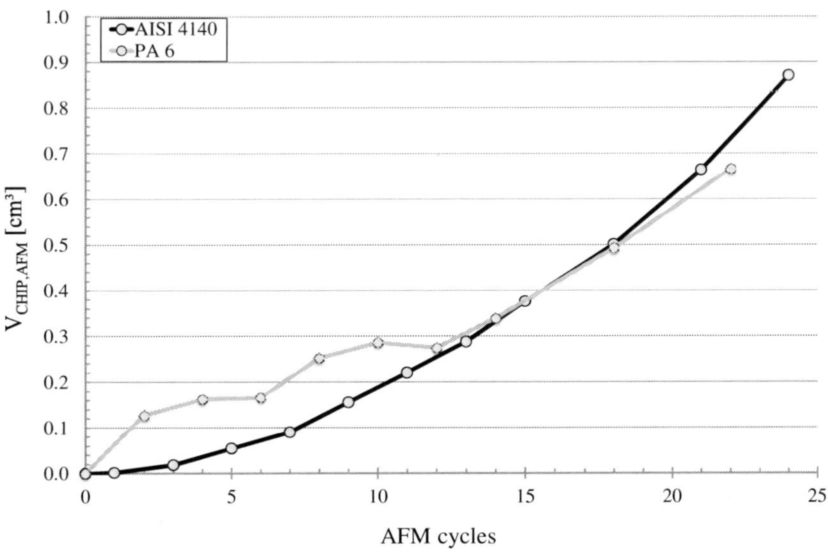

Figure 19. Removed material volume per time.

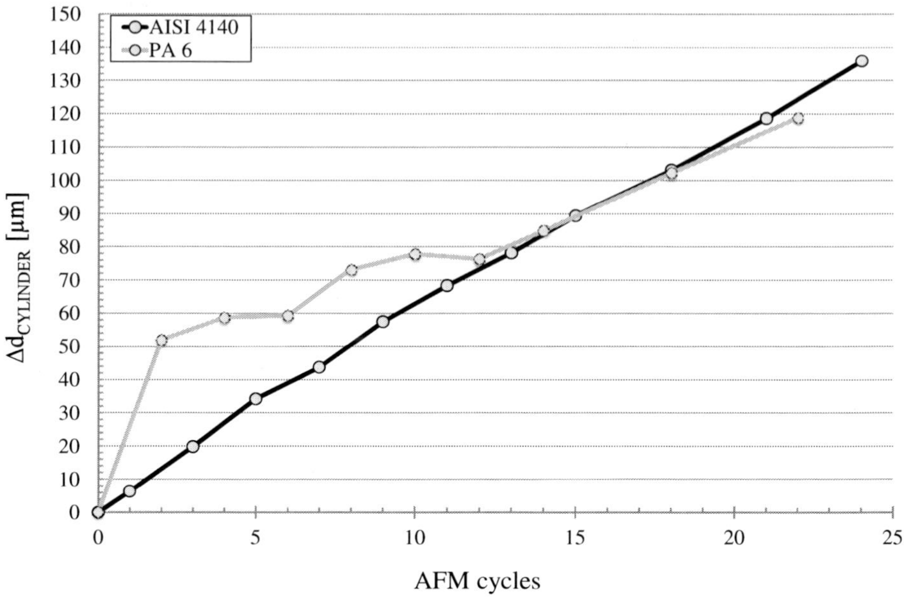

Figure 20. Increase in diameter during AFM.

Results: Roughness

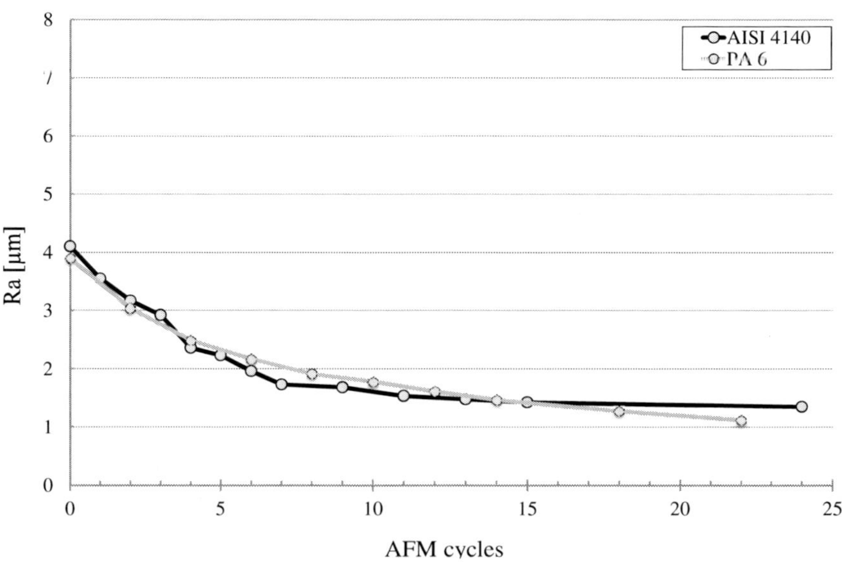

Figure 21. Mean-Ra vs. number of AFM cycles and piston pressure.

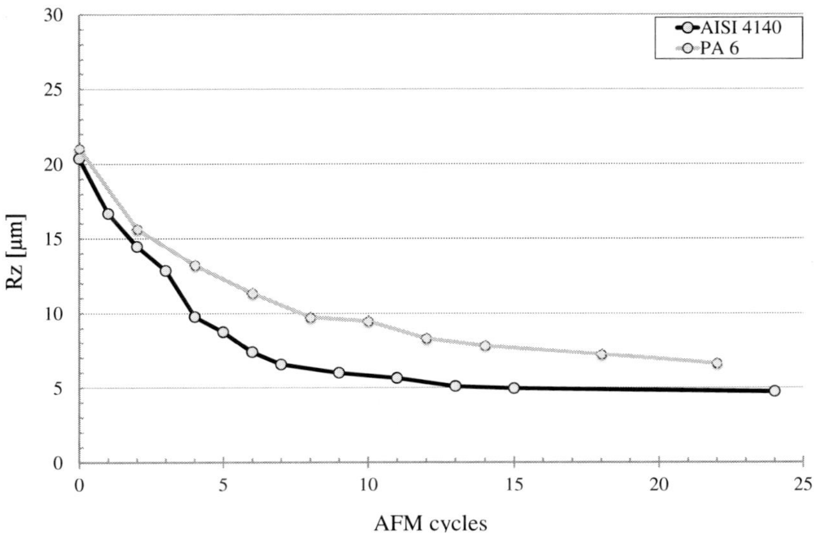

Figure 22. Mean-Rz vs. number of AFM cycles and piston pressure.

Figure 21 shows the influence of the number of cycles on the mean Ra-value for the investigated plastic components, Figure 22 shows the same for the Rz-value. In order to calculate the mean value, a total number of 20 measurements were performed for every part and AFM cycle at measurement positions X1 to X5 and 0° and 180° inside the bore.

Starting at nearly the same initial value, the steel and plastic components show a very similar behavior for the Ra-value. In contrast, the Rz-value shows measurable differences over the investigated number of AFM cycles. Here, the plastic components converge slower than the steel components. Obviously, the smoothing of the plastic surface is not as uniform as for the steel surface. Overall, a comparable surface finish can be achieved for material types at the same piston pressure level.

3.2. In-Process Measurement of Axial Forces

3.2.1. Work Pieces

The investigated work pieces are illustrated in Figure 23 and consisted of an adapter and an end piece which was put on the outlet end of the adapter. The end pieces were fixed to the adapter with mounting screws. The adapter was used in three different variants of the passage length (L_P), i.e. 50 mm, 100 mm and

150 mm. All three variants were made of machining steel AISI 12L13 and had a bore diameter of 10 mm. The inlet of the adapters directly contacted the clamping block and was therefore designed according to the diameter of the flexible clamping block. The internal geometry of the inlet was realized as a cone and provided a smooth transition of the medium into the work piece. For the end pieces, five variants with differing geometries and outlet diameters were chosen in order to represent realistic characteristics of parts that can be machined with AFM. Each end piece had a certain geometry combined of an outlet diameter (d) and a cone angle (C). The end pieces had been pre-machined by drilling and countersinking. The material of the end pieces was AISI 4140.

End piece variant	D1	D2	D3	C1	C2
Diameter d [mm]	10.0	8.7	7.1	7.1	7.1
Cone angle C [°]	0	0	0	60	30

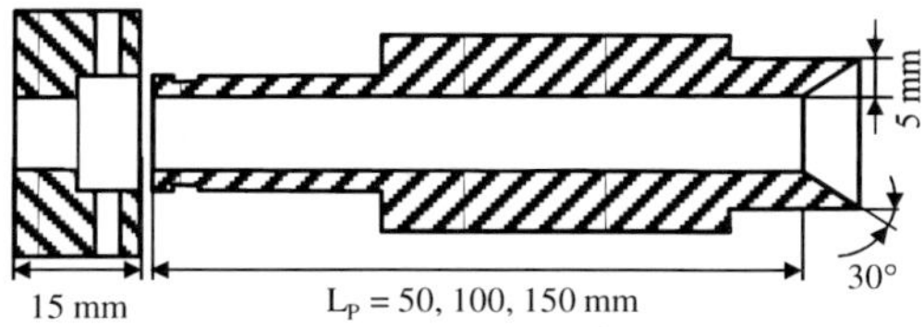

Figure 23. Work pieces for the force measurement experiments.

3.2.2. *Measurement Technique*

In advance to the present investigation some effort has been made to identify a suitable in-process force measurement setup. The commercially available one-component force sensor FUTEK LTH350 was chosen to measure the axial force component and its sensitivity to variations in parameters. The sensor had a range of +/- 2,224 N and a maximum non-linearity deviation of +/- 0.5 %. This is equal to a maximum non-linearity error of approximately 11 N. The sensor itself was not in contact with the abrasive AFM medium. In order to record the force signal sufficiently fast and precise, an HBM Gen5i data acquisition system was used. It integrates an industrial PC, data acquisition and transient recording system in one portable device. The sampling of the force signal was performed with 1 kS/s with a resolution of 16 bit by using the plug-in isolated amplifier card.

3.2.3. *Experimental Setup with Force Sensor*

The experimental setup for in-process measurement of axial forces used the same clamping block and spacer as were used for the parameter study. In order to ensure the tightness of the junction and that the adapter was freely mounted in the X-axis, there was a clearance fit between the clamping block and the adapter. The fitting surface was cleaned after each experiment to avoid the adapter to get seized in the clamping block due to leaking AFM medium and abrasives. The force sensor was put on the adapter and both were fixed with the clamping that also served as a spacer and thus provided the same conditions in all experiments. Due to the axial forces working at the adapter and the end piece during the machining, the force sensor was pushed against the clamping and detected the axial force component. The setup with the clamping block, the adapter and the force sensor is depicted in Figure 24.

Forces at work during AFM

F_{AX}	- Measured axial force
F_C	- Cone force
$F_F = F_{F,L} + F_{F,C}$	- Friction force
$F_{F,L}$	- Friction force along the lateral surface of the passage
$F_{F,C}$	- Friction force along the lateral surface of the inlet cone

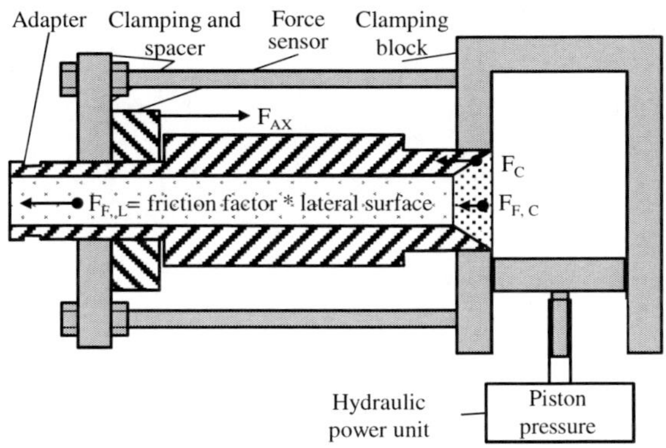

Figure 24. Experimental setup for in-process measurement of axial forces.

3.2.4. Force Model

In the experiments different axial force components worked at the work piece. With the measurement of the axial force F_{AX} under different conditions and the knowledge of the adapter geometry, the force components were analytically separated and determined. In Figure 24 typical signals of the force measurement during one AFM cycle are displayed. The curves show the increase of the axial force at the beginning of the AFM process when the AFM medium starts to enter the work piece. This phase lasts only about 1 or 2 seconds. Then a short overshoot of the force signal follows before the signal reaches a steady state. Within the steady state the force signal only fulfills a slight rise to its maximum and slightly slopes towards the end of the process. At the end of the process, when the entire medium has passed through the work piece, the force signal slopes steeply. For the evaluation of the force signal, the initial phase and the final phase have not been taken into consideration. All further results of measured forces refer to the average value of the steady state of the respective force signal.

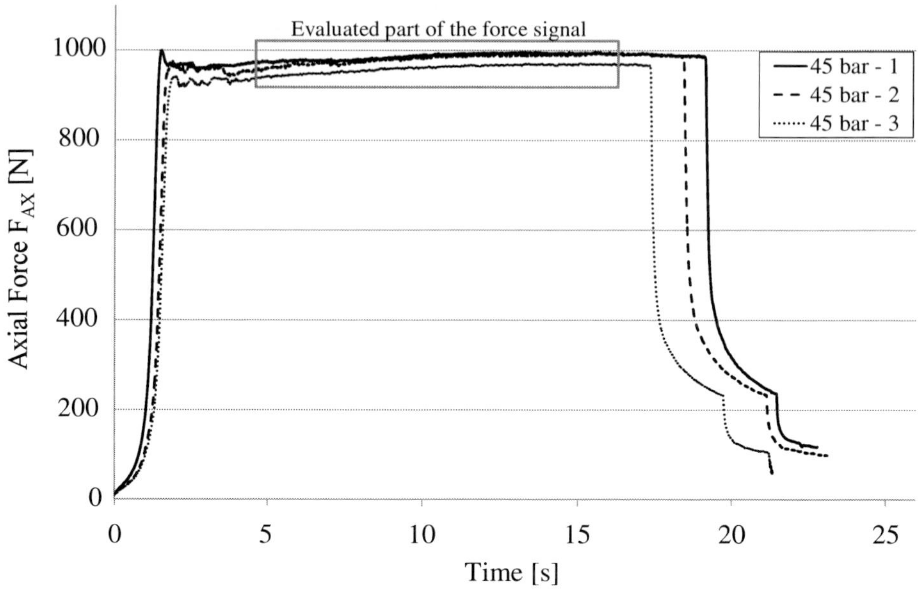

Figure 25. Evaluation of the force signal for the adapter with L_P = 150 mm without end pieces.

At the work piece, there was a balance of axial working forces that comprised the friction force (F_F), the axial force working at the inlet cone of the adapter (F_C) and the counteracting force (F_{AX}) that was measured in the experiments. The cone force (F_C) represents the axial force component of the inlet cone resulting from the narrowing of the diameter. The balance of forces can be described with Equation 2. Herein, the friction force is further divided and attributed to the friction force working at the lateral surface along the length of the passage ($F_{F,L}$) and the friction force working along the lateral surface of the inlet cone ($F_{F,C}$).

$$F_{AX} = F_F + F_C = (F_{F,L} + F_{F,C}) + F_C \quad (2)$$

The amount of the single force components can be investigated by varying the passage length of the adapter with all other parameters kept constant. Thus, a change in the measured axial force F_{AX} can be directly attributed to the change in the passage length. The friction force at the inlet cone ($F_{F,C}$) and the cone force (F_C) remain unaffected. Hence, the measured delta in F_{AX} allows for calculating the friction force along the lateral surface of the passage ($F_{F,L}$). When relating the friction force ($F_{F,L}$) to the increase of the respective surface of the adapter (A_{LP}), a friction factor (f) per lateral surface of the passage can be obtained, Equation 3.

$$f = \frac{F_{F,L}}{A_{LP}} \quad \left[\frac{N}{mm^2}\right] \tag{3}$$

When the friction factor is transferred to the lateral surface of the inlet cone (A_C), also the second component of the friction force ($F_{F,C}$) can be calculated. Eventually, the cone force (F_C) can be derived as the difference of the known axial force (F_{AX}) and friction force (F_F). Consequently, all force components can be determined.

3.2.5. Results and Discussion

In order to draw conclusions from the force measurement of F_{AX} regarding the force components, some basic investigations were performed. First, the passage length of the adapter has been varied between 50 mm, 100 mm and 150 mm with the piston pressure kept constant at 40 bar. The results are shown in Figure 26.

The results of the varied passage length show a linear response in the measured F_{AX} value. With a linear regression, the slope yields an increase of F_{AX} of approximately 260 N per 50 mm. The coefficient of determination (R^2) for the regression gives a certainty of 100 %.

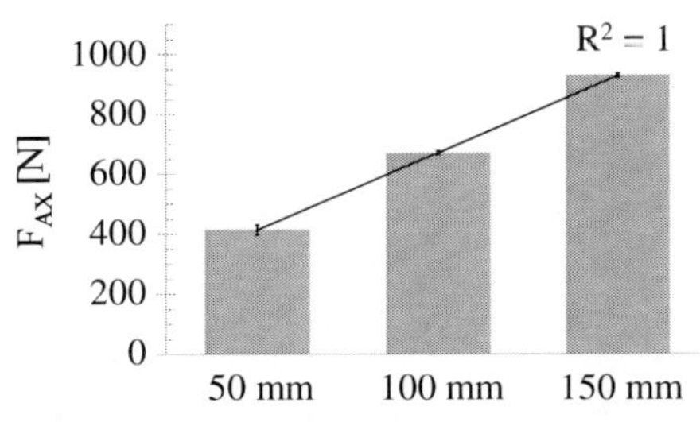

Figure 26. Influence of the increase in passage length of the adapter for piston pressure at 40 bar.

As only the passage length was increased, this 260 N can be entirely attributed to $F_{F,L}$. Given the diameter (d = 10 mm) and the length (L_P = 50 mm) of the passage, the friction factor per lateral surface can be calculated to 0.17 N/mm^2. If the friction factor is transferred to the lateral surface of the cone, which was calculated by using the outer and inner diameter of the cone section, the second axial force component $F_{F,C}$ can also be derived. $F_{F,C}$ was equally about 78 N for all the adapters. The results for the calculation of the force components are summarized in Table 2. Following Equation 2, also F_C can be concluded. It is small compared to F_F.

Table 2. Analytical calculation of forces at work.

L_P		[mm]	50	100	150
F_{AX}	Measured axial force	[N]	413	671	929
$F_{F,L}$	Calculated friction force, passage length	[N]	262	523	785
$F_{F,C}$	Calculated friction force, cone section	[N]	78	78	78
$F_F = F_{F,L} + F_{F,C}$	Calculated friction force	[N]	340	601	863
$F_C = F_{AX} - F_F$	Calculated cone force	[N]	73	70	66

Second, the influence of the adjusted piston pressure on the axial force was investigated. Three different piston pressures of 35 bar, 40 bar and 45 bar were applied for the adapter with an L_P of 150 mm. From Figure 27 the results can be seen.

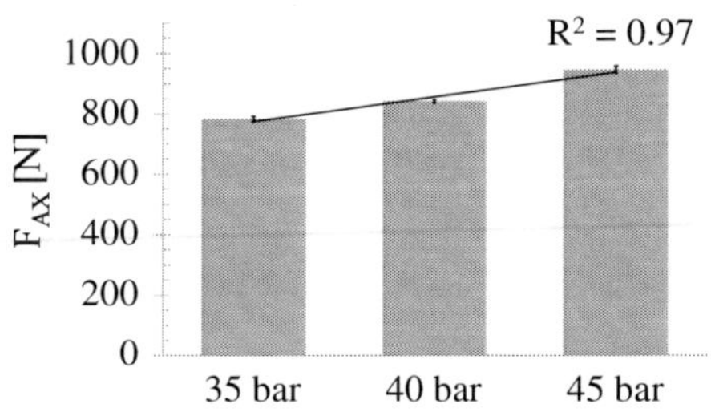

Figure 27. Influence of the piston pressure level on F_{AX} for the adapter with L_P = 150 mm.

As Figure 27 shows, the increase in the piston pressure also results in an increase in the measured F_{AX}. The slope yields an increase of approximately 19.3 N per bar pressure. When comparing absolute values of different measurement trials, this value becomes important. The coefficient of determination for the regression is $R^2 = 97$ %. The pressure adjustment could be performed with an accuracy of about +/- 2 bar, i.e. +/- 38.6 N. However, this adjustment error becomes negligible when comparing measurements with the same adjustment of the piston pressure. Variation of the medium temperature is another influencing parameter and an additional source of uncertainty, but it was found to be negligible as well, when using the same adapter for all end pieces.

Finally, the five end pieces were mounted to the adapter with L_P = 150 mm to identify the influence of the different inner passage geometries on the value of the measured F_{AX}. The results are summarized in Figure 28.

The bar labeled with "W/O" indicates F_{AX} for only the adapter without end piece. It was used as reference measurement. End piece D1 was only an extension of the adapter passage with the same diameter of 10 mm. Interestingly, F_{AX} mainly increases with the passage length by approximately 6.3 N/mm. The narrowing of the diameter in end pieces D2 and D3 as well as the different cone angles of end pieces C1 and C2 only account for an additional 18 to 21 N. As a consequence it can be concluded that the length of the additional passage length is the main influence on F_{AX}.

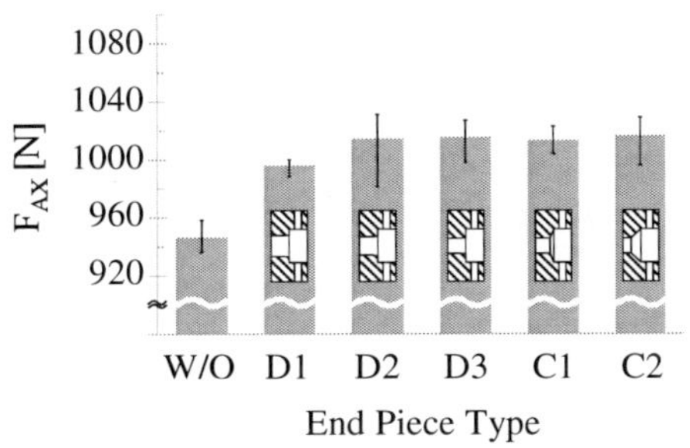

Figure 28. Influence of the end piece geometries on the F_{AX} for piston pressure at 45 bar.

CONCLUSION

This chapter presented recent research results in one-way AFM. First, a parameter study was conducted in order to evaluate the influence of piston pressure and number of AFM cycles on form tolerance, surface quality and wear. For the used steel AISI 4140 it could be shown that the diameter of the machined bore is linearly increasing between 3.26 μm/cycle and 6.16 μm/cycle for 40 bar or 70 bar, respectively. AFM at 70 bar removes nearly twice as much material per cycle than AFM at 40 bar. In order to compare the material removal rate with conventional cutting processes, $V_{CHIP, AFM}$ as removed volume per AFM cycle was introduced and calculated for different pressure levels. The increase could be described as a quadratic function for all four considered pressure levels.

Furthermore, the diameter did not increase homogeneously over the length of the bore but instead a so-called bell-mouthing effect could be observed towards the side of the medium outlet. It can be avoided by applying appropriate fixtures and adapters depending on the machining task. The out-of-roundness error increases with increasing piston pressure and so does the out-of-straightness error. If the form tolerance of the treated components is critical, small piston pressures should be applied. However, higher piston pressures will lead to much faster improvement of the surface quality along with a significant faster machining time. At 70 bar, Ra and Rz show already after 7 AFM cycles only negligible improvement potential. Beyond that, this pressure level shows the tendency of decreasing Ra- and Rz-values with further AFM. After 42 cycles those values have been steadily increased whereas the 40 bar pressure level has not even reached a minimum value yet. Therefore, an optimization in process control was suggested which can be compared with a roughing and finishing procedure in conventional machining. Combining AFM with 70 bar for the first 7 cycles and 40 bar for another 16 cycles revealed the potential to theoretically reduce the number of AFM cycles by 45 % and moreover to reduce the machining time by as much as 58 % given that a minimum possible Ra- or Rz-value is reached, respectively. These findings have to be verified with further long-term investigations, but the potential is obvious and should be exploited. In case of the plastic material PA 6, the removed volume per AFM cycles as well as the widening in diameter was comparable to the found results for AISI 4140. However, a tendency of a decreasing removal rate could be observed for PA 6 at the end of the acquired AFM cycles. Future long-term investigations should be performed to prove this. The measured Ra-values showed a comparable behavior to the steel components whereas the Rz-values were consistently higher indicating a less uniform smoothing in the case of PA 6 (smaller reduction of peaks and valleys in the corresponding surface profile). As a matter of fact, PA 6 has advantages over the steel in terms of purchase price, machinability and corrosion resistance. Therefore, it can be recommended as material for the construction of wear components for clamping and adapting purposes.

Second, a novel in-process axial force measurement setup for the determination of axial forces was presented. A model for the equilibrium of forces was introduced. Basic investigations under the variation of piston pressure and passage length (i.e. the machined surface) allowed for the quantification of the different force components. The results showed that the axial force F_{AX} increased linearly with a gradient of 19.3 N per bar piston pressure. In the same way, F_{AX} increased linearly with increasing passage length. An additional passage length of 50 mm yielded an additional axial force of 260 N. By relating the friction force component of the measured axial force to the machined surface of the work piece, a friction factor was obtained. Further investigation of end pieces with different diameters and geometries showed that the machined surface is the dominating influence on the axial force. In future work the results from the in-process force measurement like the friction factor per machined surface are to be applied in numeric simulation models of the AFM process. Further, the in-process force measurement can be used for controlling the AFM process. Therefore, further investigations have to be performed in order to determine the influence of piston pressure, change of the machined surface (i.e. widening of the diameter, improvement of the surface) and medium temperature on the measured axial force.

References

[1] Kloos, K.H., Macherauch, E. (1987). Development of Mechanical Surface Strengthening Processes from the Beginning until Today. In: H. Wohlfahrt, R. Kopp, O. Vöhringer, *Shot Peening, Science – Technology – Application* (pp. 3-28), Oberursel: DGM Informationsgesellschaft.

[2] Extrude Hone (2012). *http://www.extrudehone.com/products/media/*, retrieved 05.12.2012.

[3] Przyklenk. K. (1987). Druckfließläppen – Feinbearbeitung mit Schleifpasten. In: Saljé, E. (Ed.), *Feinbearbeitung. Schlüsseltechnologie für die Fabrik der Zukunft* (9.1-9.18). Braunschweig, Germany.

[4] Sankar, M. R., Jain, V. K., Ramkumar, J. (2007). Abrasive flow machining (AFM): An Overview. *http://www.psgtech.edu/psgias/smart_machine_tools/V.K.Jain.pdf*, 23.11.2012.

[5] Jain, V. K., Adsul, S. G. (2000). Experimental investigations into abrasive flow machining (AFM). *International Journal of Machine Tools & Manufacture*, 40, 1003-1021.

[6] Rhoades, L. J. (1989). *Abrasive Flow Machining. Technical Paper, Society of Manufacturing Engineers* (SME). MR89-145.

[7] Singh, S., Shan, H. S., Kumar, P. (2002). Wear behavior of materials in magnetically assisted abrasive flow machining. *Journal of Materials Processing Technology*, 128, 155-161.

[8] Uhlmann, E., Szulczynski, H. (2005). Precise Finishing of Inner Contours with Abrasive Flow Machining. *International Journal for Manufacturing Science & Technology*, 7, 33-39.

[9] Szulczynski, H. (2007). *Verfahrensgrundlagen und Technologie des Hubschleifens mit viskosen Schleifmedien.* PhD Thesis, Stuttgart: Fraunhofer IRB.

[10] Kar, K. K., Ravikumar, N. L., Piyushkumar B. T., Ramkumar, J., Sathiyamoorthy (2009). Performance evaluation and rheological characterization of newly developed butyl rubber based media for abrasive flow machining process. *Journal of Materials Processing Technology*, 209, 2212-2221.

[11] Fang, L., Zhao, J., Sun, K., Zheng, D., Ma, D. (2009). Temperature as sensitive monitor of efficiency of work in abrasive flow machining. *Wear*, 266, 678-687.

[12] Uhlmann, E., Mihotovic, V., Szulczynski, H., Kretzschmar, M. (2010). Developing a Process Model for Abrasive Flow Machining. In: Aurich, J. C., Dornfeld, D. (Eds.), *Burrs – Analysis, Control and Removal.* Berlin, Germany: Springer.

[13] Baehre, D., Bruennet, H, Swat, M. (2012). Investigation of one way abrasive flow machining and in process measurement of axial forces. In: Wegener, K. (Ed.), *Proceedings of the 5th CIRP International Conference on High Performance Cutting* (pp. 419-424). Zuerich: Elsevier.

[14] Uhlmann, E., Mihotovic, V., Coenen, A. (2009). Modelling the abrasive flow machining process on advanced ceramic materials. *Journal of Materials Processing Technology,* 2009, 6062-6066.

[15] Rhoades L. J. (1993). Abrasive Flow Machining – Progress in Productivity. *Technical Paper, Society of Manufacturing Engineers* (SME). MR93-129.

[16] Williams, R. E., Rajurkar, K. P., Rhoades, L. J. (1989). *Performance Characteristics of Abrasive Flow Machining. Technical Paper, Society of Manufacturing Engineers* (SME). FC89-806.

[17] Perry, W. B. (1975). Properties and Capabilities of low-pressure abrasive flow mediae. *Technical Paper, Society of Manufacturing Engineers* (SME). MR75-831.

[18] Jain, R. K., Jain, V. K., Dixit, P. M. (1999). Modeling of material removal and surface roughness in abrasive flow machining process. International *Journal of Machine Tools & Manufacture*, 39, 1903-1923.

[19] Borchers, J. (1980). Why? Abrasive flow deburring. *Technical Paper, Society of Manufacturing Engineers* (SME). MR80-101.

[20] Gorana, V. K., Jain, V. K., Lal, G. K. (2004). Experimental investigation into cutting forces and active grain density during abrasive flow machining. International *Journal of Machine Tools & Manufacture*, 44, 201-211.

[21] Raju, H. P., Narayanasamy, K., Srinivasa, Y. G., Krishnamurthy, R. (2005). Characteristics of extrude honed SG iron internal primitives. *Journal of Materials Processing Technology*, 166, 455-464.

[22] Gorana, V. K., Jain, V. K., Lal, G. K. (2006). Forces prediction during material deformation in abrasive flow machining. *Wear*, 260, 128-139.

[23] Schmitt, J., Diebels, S. (2013). Simulation of the Abrasive Flow Machining Process. Z. *Angew. Math. Mech.*, 2-3, 147-153.

[24] Hull, J. B., O'Sullivan, A. J., Fletcher, S. A., Trengove, J. M. (1992). Rheology of carrier media used in abrasive flow machining. *Key Engineering Materials*, 72-74, 617-626.

[25] Raja, J., Muralikrishnan, B., Shengyu, F. (2002). Recent Advances in Separation of Roughness, Waviness and Form. *Journal of the International Societies for Precision Engineering and Nanotechnology*, 26, 222-235.

[26] DIN EN ISO 4287 (2010). *Surface Texture: Profile Method – Terms, definitions and surface texture parameters.* Berlin: Beuth Verlag.

[27] ASME (2009). *B46.1: Surface Texture* (Surface Roughness, Waviness, and Lay). New York: ASME.

In: Recent Advances in Abrasives Research ISBN: 978-1-62808-566-2
Editor: Dirk Bähre

Chapter 8

ADVANCES IN MACHINING WEAR-RESISTANT COATINGS BY GRINDING AND HONING

Sebastian Goeke*[*]*, Sascha Rausch, Petra Kersting and Dirk Biermann

Institute of Machining Technology, TU Dortmund University, Dortmund, Germany

ABSTRACT

Wear-resistant coatings are used to increase the lifetime of tribologically stressed workpieces. In this paper, different process strategies for finishing and grinding of hard material coatings on various workpiece shapes are investigated. Both processes were characterized regarding the achievable surface quality and the flexibility concerning the shape of the coated specimens in a comparative analysis. The main influencing variables, e.g., process forces, machine time, and tool conditions were analyzed for both processes. With adapted process strategies it is possible to reduce the surface roughness of Rz = 10,...,30 μm to a mirror-like surface quality with a surface roughness of Rz < 0.3 μm. By grinding on machining centers, also free-formed surfaces can be ground.

Keywords: Grinding, honing, finishing, coatings

---

[*] Corresponding author. Email: goeke@isf.de.

1. INTRODUCTION

Thermally sprayed coatings based on tungsten carbide are a widely used solution to enhance the wear resistance and lifetime of functional surfaces [1]. The improved resistance against abrasive, corrosive, and cavitative wear leads to an advantageous behavior under tribological stress [2, 3]. Typical industrial applications like turbines, landing gears, combustion engines, hydraulic modules, dies and molds illustrate the importance of thermally sprayed coatings. In addition, high surface qualities and shape accuracies are needed to provide the continuously increasing demands on these surfaces. With respect to the high surface roughness resulting from the spraying process, a subsequent abrasive machining is necessary in order to improve the shape accuracy and surface quality. The characteristics of the coated surfaces can be improved by finishing operations [4]. Due to the exceedingly high hardness of the coating, the machining of these coatings by milling or turning processes is not favorable. Abrasive processes with geometrically undefined cutting edges are rather suitable for the machining of these materials. Both, grinding and honing processes are commonly used to generate a smooth surface finish. The kinematics of the honing process and the use of diamond as hard cutting material allow the generation of surface qualities with a significantly reduced surface roughness. Depending on the process strategy and with regard to the specific application, either smooth surface topographies with a minimized surface roughness or plateau-like bearing surfaces can be achieved. The tribological behavior, especially the frictional response, of those structured surfaces is significantly improved [5]. During their operational life, under mixed and boundary lubrication conditions, plateaued surfaces are characterized by a shorter running-in period than non-plateaued surfaces. Additionally, the steady state wear rate is lower than when using non-plateaued surfaces [6]. The tribological behavior of tungsten carbide coatings machined with different surface finishing processes like grinding and polishing in comparison with wire electrical discharge machining (EDM) for example showed that a smooth surface causes a significantly lower wear rate as well as a shortening of the running-in period [7]. In order to combine the positive aspects of a highly wear-resistant coating with a surface structuring, this study aims at evaluating the options to machine these coatings by microfinishing and grinding.

For the machining of complex free-formed workpieces such as forming tools, grinding on machining centers using cylindrically and spherically shaped grinding tools, so called mounted points, are suitable [8, 9]. Resulting from the low spindle speed of the machining center and the small tool diameter, the cutting velocity is strongly limited. In combination with a small depth of cut, limited by the bond strength of the tool, micro ploughing occurs in the contact zone. Additionally, the feed speed has to be decreased to achieve an adequate chip thickness [10, 11]. The conventional high flow rate of lubricant, commonly needed for grinding processes, can be reduced while the cutting speed is at a low level. Vitrified bonded grinding tools provide a suitable coolant supply to the grinding arc that leads to a reduction of residual stresses in the workpiece surface [12]. The achievable roughness values are strongly influenced by the dressing conditions of the grinding tools and can be improved by using smaller grains [13]. In case of dry dressing, the temperature occurring in the contact zone between dressing and grinding wheel shows a linear dependency of the depth of dressing cut. Consequently, the depth of dressing cut has to remain at a low level, thereby the diamond grains won't be damaged [14]. In addition to this, the relative velocity of the dressing and the grinding wheel has generally a high impact on the surface roughness. The roughness is mainly influenced by the dressing parameter for small values of material removal [15].

In this paper, the grinding and mircofinishing of tungsten carbide coatings are presented. In fundamental investigations the machined surfaces were analyzed by means of optical and tactile measurement systems. In order to give an outlook for further necessary research, various possible applications concerning the shape of the specimens are shown.

2. Fundamental Investigations

The hard material coatings used in the investigations are based on tungsten carbide in a cobalt/chrome binder matrix. They were deposited by High Velocity Oxygen Fuel (HVOF) and warm spraying. During the spraying the feedstock powder with a particle size of 2,...,10 μm is heated and accelerated in a combustion process. Based on the high particle velocity and the small grain size, a lamellar layer structure with a low porosity, high bonding strength, and superior wear resistance is generated [2, 3]. Due to the high hardness of approximately 1300 HV0.3, the machining processes were carried out using diamond as cutting material. To characterize both processes and analyze them, different influencing variables like infeed force, process time, and tool type (e.g., grain size, bonding) were varied.

The generated surfaces have been analyzed using a tactile measurement system *MahrSurf XR 20* and a confocal white light microscope *Nanofocus μsurf*. The recorded area data enables the evaluation of 3D models in order to get a visual impression of the surface topography. By extracting profiles of the area data, the tactile roughness measurements could be validated. In addition to the maximum height of profile Rz, the peak height and the valley depth of the profile were measured and analyzed in form of a bearing curve. These parameters are suitable to describe precisely the expected functional behavior of the surface topography. The maximum height of profile Rz is one of the most widely used roughness parameters in industrial companies [16].

2.1. Microfinishing

The short-stroke honing - also called microfinishing, superfinishing or simply finishing - is characterized by the superposition of two cutting motions, the rotating workpiece and the oscillating finishing tool. For the machining of the cylindrical specimens, the finishing machine *Supfina 202* was adapted on a conventional lathe *Boehringer V800*. Short-stroke honing processes are conducted with honing stones or with belts alternatively. Due to the constant contact between honing stone and workpiece, abrasive particles might remain in the contact area. As shown in earlier studies, various parameters like the quantity of cooling lubricant, the contact pressure as well as the oscillation frequency influence the loading on the stone surface and thereby the wear of the honing stone. Depending on the special stone characteristics like grain type, grain size and bonding, it is necessary to determine optimal values for these parameters [17, 18]. By using finishing films, as done in this study and shown in Figure 1, several disadvantages like an unpredictable tool wear or wear particles which interfere the process behavior can be avoided.

With regard to the used grain size, high material removal rates can be achieved. Similar to honing stones, every finishing belt specification has its own steady state roughness value, which can be generated [19, 20, 21]. By using larger grain sizes for the premachining process, a comparatively high removal rate can be achieved, but a higher surface roughness is generated. To combine the conflicting demands of a high removal rate and a smooth surface finish, the finishing process is conducted in several subsequent finishing steps with varying film specifications.

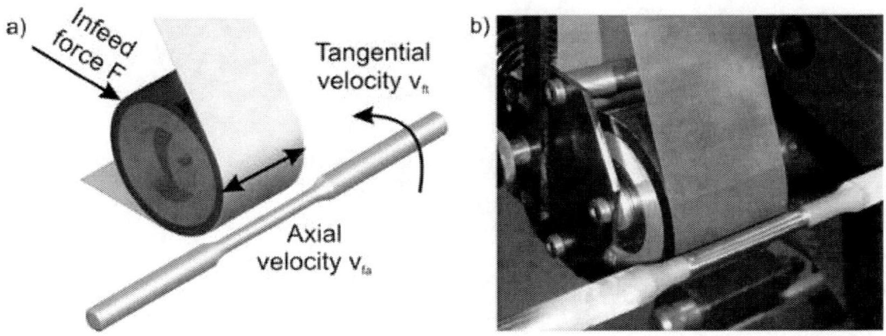

Figure 1. Experimental setup – Microfinishing. (a) Model and Kinematics. (b) Real process.

In order to increase the resulting surface quality, oil was used as cooling fluid with minimum quantity lubrication (MQL) instead of flooding. Earlier studies about the influence of lubrication conditions have shown that a reduced cooling supply can lead to advantageous machining conditions.

Using MQL during the micrcofinishing process, results in the generation of a kind of slurry in the contact zone. This slurry, consisting of grains, microchips, and oil, leads to an additional lapping phenomenon with an improvement of the surface quality [22]. In contrast, conducting the process without any cooling lubricant can result in a thermally affected layer on the surface which is characterized by tensile stresses [23]. Depending on the desired surface topography of the workpiece, different kinds of finishing films can be used to influence the surface regarding the material ratio. Electrostatically adjusted finishing films are characterized by an aggressive machining behavior. Based on the adjustement process, a high grain protrusion is generated leading to an increased material removal rate. Due to this, they are usually used for the premachining in combination with larger grain sizes. In order to reach a smooth surface finish in the end, the final finishing steps are carried out with muddled finishing films. In this case the grains are randomly distributed on the polyester layer in a resinoid bonding. Both types of finishing films are shown in Figure 2. The finishing process was conducted in three process steps with subsequently smaller grain sizes ranging from $d_K = 125$ μm to $d_K = 9$ μm.

Additionally, the influencing process parameters infeed force F and process time t were analyzed using statistical methods from the design and analysis of computer experiments [24] based on a combination of the central composite design and the Box-Behnken design.

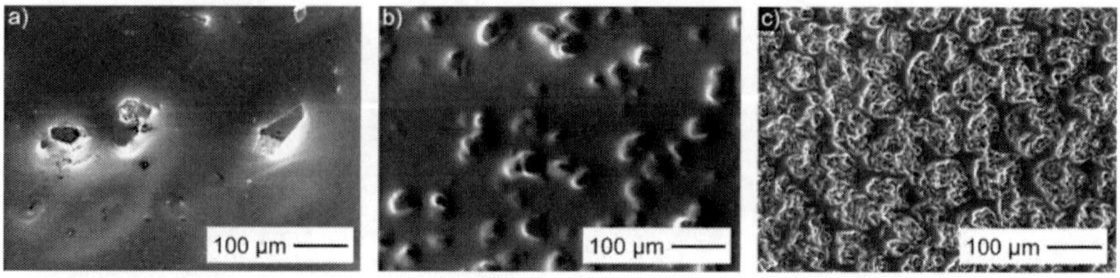

Figure 2. Scanning Electron Microscope pictures of the finishing films (20 kV). (a) Elecrostatically adjusted grains with a size of $d_K = 125$ µm. (b) Loosely distributed grains with a size of $d_K = 30$ µm and (c) $d_K = 9$ µm.

To ensure the reliability of the empirical model, the center point was repeated five times. By using three different pressure rollers, the relation between the infeed force and the resulting contact pressure was measured. The infeed force varied between $F = 60,...,600$ N, while the process time ranged between $t = 10,...,60$ s. By using an electrostatically adjusted finishing film with a grain size of $d_K = 125$ µm for the premachining it is possible to generate surface qualities characterized by a maximum height of profile between Rz = 1.6,...,15 µm. Choosing an infeed force higher than $F = 300$ N the as-sprayed surface roughness of Rz = 10,...,30 µm can be decreased within $t = 30$ s to a surface roughness of Rz < 2 µm. The roughness values were measured orthogonal to the cutting direction. As shown in Figure 3 an increased infeed force as well as a longer process time result in a lower surface roughness. In contrast to a honing process which is conducted with stones instead of films, there is no critical reversal point regarding the process efficiency when increasing the infeed force. In case of finishing with honing stones, when choosing an infeed force which is too high, the efficiency of the process decreases [17]. When using finishing films, the strength of the grain bondage is the only limitating factor. An infeed force which exceeds the maximal loading capacity will lead to a detachment of the layers of the finishing film. By increasing the infeed force, the average contact pressure is increased as well and leads to a higher removal rate. The same effect occurs by using a harder pressure roller. Regarding the surface roughness, most studies examined a smoother surface finish when using a softer pressure roller. A decreased hardness of the pressure roller causes an increased contact area [21, 25]. Coinciding with the greater contact area, the average contact pressure is reduced, leading to a lower removal rate. With regard to the high surface roughness of the as-sprayed coating, which has to be processed, these two effects seem to negate each other for the finishing of the tungsten carbide coating. The hardness of the pressure roller has only a small effect on the generated surface roughness compared with the influence of the infeed force and the process time.

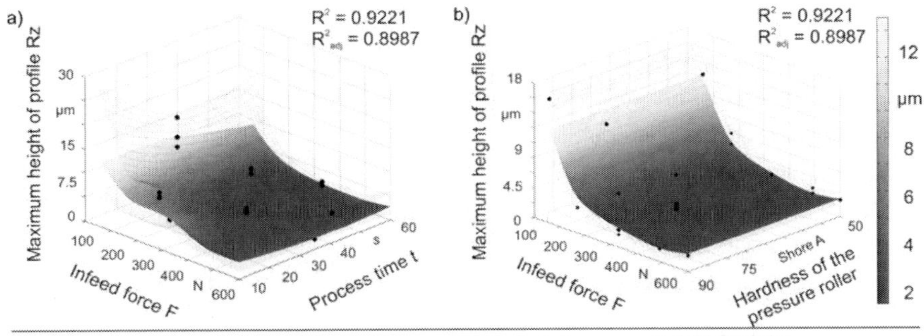

Material:	86WC-10Co4Cr	Cutting speed:	v_c = 21.2 m/min
Grain size:	d_K = 125 µm	Oscillation width:	l_{os} = 3 mm
Grain type:	Diamond	Oscillation frequency:	f_{os} = 21 Hz
Coolant:	Oil	Belt feed:	v_{fb} = 25.6 mm/min

Figure 3. Maximum height of profile depending (a) on the infeed force and the process time respectively (b) on the infeed force and the hardness of the pressure roller. The determination coefficients are based on a leave-one-out-cross-validation.

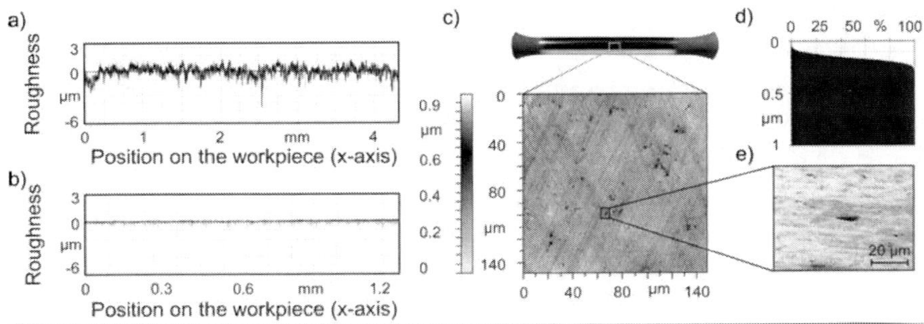

Material:	86WC-10Co4Cr	Cutting speed:	v_c = 21.2 m/min
Grain size:	d_K = 125; 30; 9 µm	Oscillation width:	l_{os} = 3 mm
Grain type:	Diamond	Oscillation frequency:	f_{os} = 21 Hz
Coolant:	Oil	Pressure roller:	90; 75 Shore A
Infeed force:	F = 270...330 N	Process time:	t = 35 s; t_{ges} = 105 s

Figure 4. Surface quality of the coated workpiece after different finishing steps. (a) Roughness profile after finishing with a grain size of d_K = 125 µm. (b) Roughness profile after using three finishing steps with subsequently smaller grain size d_K = 125,...,9 µm. (c) Surface topography measured with an optical whitelight microscope. (d) Material ratio curve. (e) SEM of the inherent porosity of the coating.

Conducting two additional process steps with subsequently smaller grain sizes of $d_K = 30$ µm and $d_K = 9$ µm enables a significant decrease of the surface roughness. When using finishing films with a grain size of $d_K = 30$ µm, the surface roughness can be decreased to a maximum height of profile Rz < 0.3 µm (Figure 4). As an overall result, within an entire process time of $t = 90$ s, mirror-like surface qualities can be generated without using time- and cost-affordable polishing and lapping processes [26]. Any further improvement of the surface quality by using smaller grains is limited due to the inherent porosity of the coating. In combination with the fact that the powder particles have a size of 2,...,10 μm it is not possible to get a smoother finish than Rz = 0.2,...,0.3 μm in three subsequent process steps. In contrast, the porosity can function as an oil support volume. Very smooth surfaces without any surface structuring usually show poor wear behavior. A sufficient oil holding capacity is necessary to provide enough lubrication in the contact zone [27].

2.2. Grinding of Hard Material Coatings

The use of cubical boron nitride (CBN) as cutting material can lead to small cracks in the ground surface, in case of applying wrong dressing conditions. Because of the higher hardness of diamond grains in comparison to the tungsten carbide coating, the surface can be ground without occurring defects [9, 28, 29]. Vitrified bonded grinding tools with diamond grain sizes of 46 and 126 μm, according to *Federation of European Producers of Abrasives* (FEPA) standard, were applied to get a smooth surface finish. The porose topography of the bonding supports good coolant supply and chip removal. The roughness values can be reduced by the use of smaller grains with an increased grain density. Subsequently, a grinding tool with a rubber bonding and smaller diamond grains was used in order to polish the workpiece surface. In Figure 5 scanning electron microscope pictures of the topographies of the applied grinding tools are depicted.

The grinding tools were prepared on the five axis dressing machine *Geiger AP-800 Fusion* without using coolant (Figure 6). During the investigation the tools have been dressed prior to each experiment with a total dressing infeed of $a_{ed} = 20$ μm in steps of 4 μm per pass and an oscillation velocity of $f_d = 100$ mm/min. The dressing speed ratio was at a constant value of $q_d \approx 0.8$, resulting from the quotient of the dressing wheel velocity v_r and the grinding wheel velocity v_{sd}. Silicon carbide grinding wheels (SiC) with a grain size of C120 were applied as cutting material for the dressing process. The overlap factor was $U_d = 10$ using a dressing wheel width of $b_d = 13$ mm and a dressing traverse speed of $f_{ad} = 1.3$ mm/Revolution. The macroscopic shape accuracy of the dressed grinding wheel was measured with a stationary camera system in the machining center.

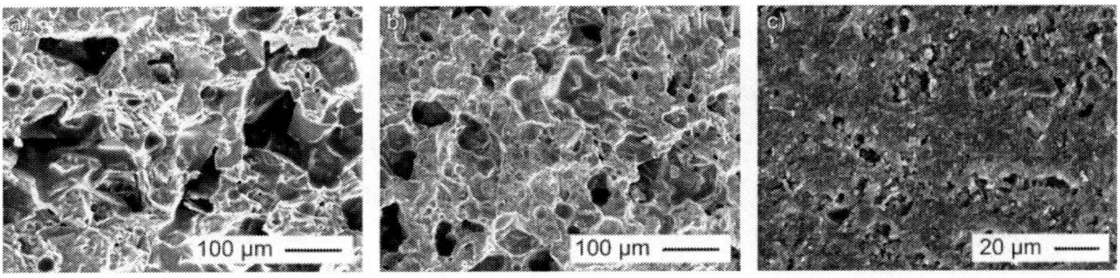

Figure 5. Scanning Electron Microscope pictures of the applied grinding tools (20 kV). (a) Vitrified bonded D126, (b) vitrified bonded D46, and (c) elastic bonded polishing grinding tool.

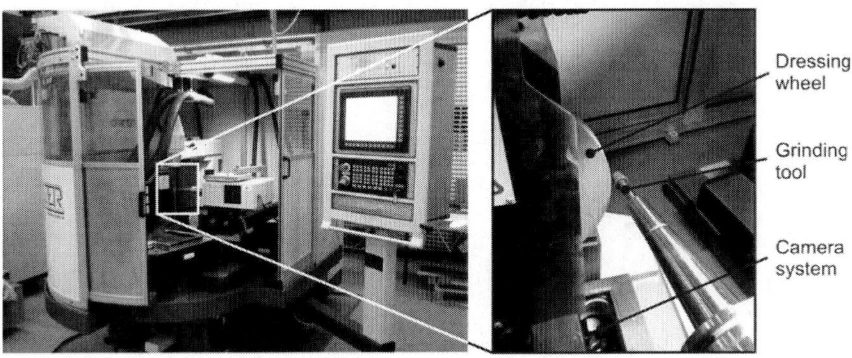

Figure 6. CNC-Dressing machine for the preparation of the grinding tools.

In order to provide a high flexible machining process, the grinding process was adopted to the conventional five axis machining center *Deckel Maho DMU50 eVolution*. The used grinding tools have a diameter of $d \approx 15$ mm. Due to the small diameter of the grinding tools and the limited spindle speed of the machining center $n_{max} = 18{,}000$ 1/min, the cutting speed was set to a constant level of $v_c = 10$ m/s. During the investigation the process parameter values feed speed $v_f = 100,...,800$ mm/min and depth of cut $a_e = 2,...,16$ µm have been varied and the width of cut was constant at $a_p = 2$ mm. The experimental setup for the grinding of flat surfaces is presented in Figure 7.

The grinding of tungsten carbide coatings was evaluated during flute grinding processes with three spark-out passes. An exemplary surface topography of the ground surface generated by the D46 mounted point is presented in Figure 8a.

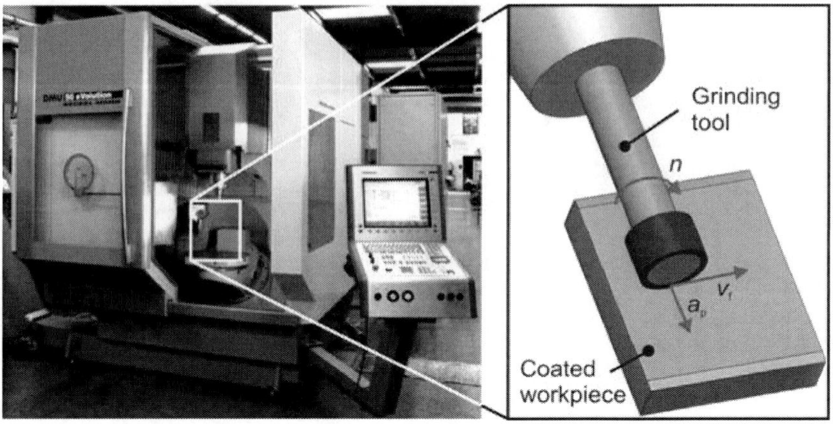

Figure 7. Experimental setup of the grinding process on machining centers using mounted points.

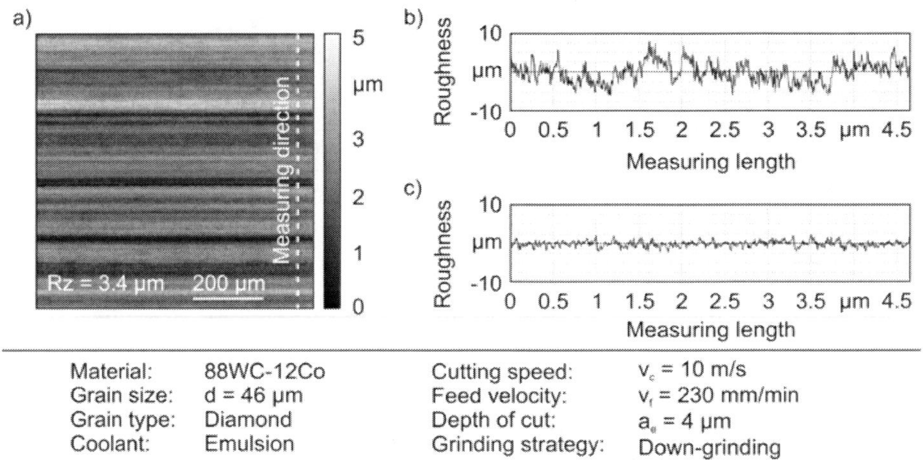

Figure 8. (a) Surface topography of the ground surface. Roughness profile of the (b) unmachined and the (c) ground WC-Co coating.

The measured roughness of this surface was Rz = 3.4 μm by applying the process parameter values depth of cut $a_e = 4$ μm and feed velocity $v_f = 230$ mm/min. Due to the low feed velocities and small tool diameter, in contrast to flat surface grinding processes, the feed per rotation can hardly be determined as single grinding grooves at the surface. This results in a reproduction of the envelope of the grinding tool topography on the ground surface. Hence, the surface roughness Rz is strongly dominated by the size and the density of the cutting material, and the dressing condition of the tool. The cross-section profile of the surface topography is depicted in Figure 8b for the as-sprayed surface and in Figure 8c for the ground surface. For the evaluation of the measured surface roughness, again DACE models have been established and calculated by the software *Matlab®* (Figure 9). For both grinding tools different tendencies were observed. In case of applying D126 mounted points, the roughness values increase nearly linear with the feed speed. The depth of cut shows no significant influence while the feed speed remains at low values. For the maximum feed speed of v_w = 800 mm a local minimum of the depth of cut can be observed at a_e = 10 mm. The range of the roughness values Rz varies between 4.2 and 10.1 μm.

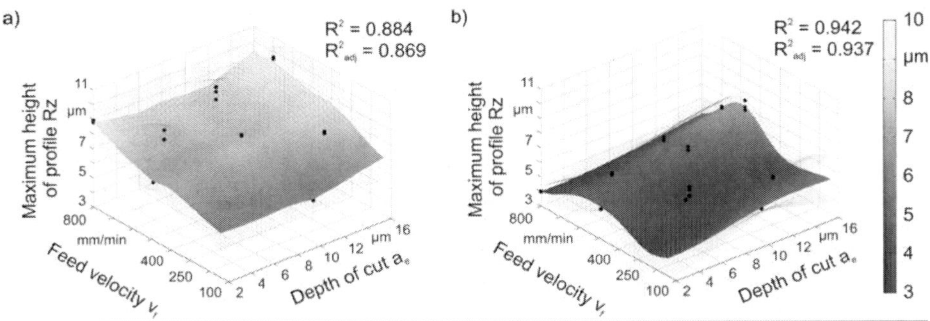

Material:	88WC-12Co	Cutting speed:	v_c = 10 m/s
Grain size:	d = 46 and 126 μm	Feed velocity:	v_f = 100...800 mm/min
Grain type:	Diamond	Depth of cut	a_e = 2...16 μm
Coolant:	Emulsion	Grinding strategy:	Down-grinding

Figure 9. Influence of the feed velocity v_f and the depth of cut a_e to the roughness Rz for vitrified (a) D126 and (b) D46 mounted points. The determination coefficients are based on a leave-one-out-cross-validation.

The results of the D46 mounted point show a slight linear increase of the roughness depending on the depth of cut. Due to the smaller grains, the roughness is reduced and the range of Rz varies between 3.4 and 7.4 μm. In contrast to the D126 tool, the roughness has a local maximum at a feed speed of $v_f \approx$ 610 mm/min. As distinguished from the microfinishing process, a defined volume of material can be removed by the grinding process.

3. Advanced Process Kinematics

3.1. Flat Surface Finishing

By adapting the finishing machine into a machining center (*Ixion TLF 1004*) optionally, the rotational motion of the workpiece can be replaced by a linear motion. The linear tangential motion enables the machining of flat surfaces and thereby increases the spectrum of workpiece shapes. As shown in Figure 10, an adapted finishing tool was used to achieve the experimental results. The adapted tool construction is able to compensate deviations regarding the angle between workpiece and finishing tool. This ensures a reliable machining of the whole surface.

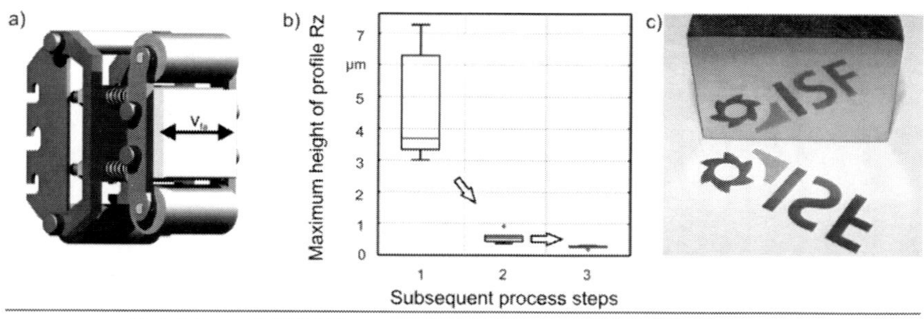

Material:	88WC-12Co	Cutting speed:	v_c = 6.3 m/min
Grain size:	d_K = 125; 30; 9 µm	Oscillation width:	l_{os} = 3 mm
Grain type:	Diamond	Oscillation frequency:	f_{os} = 21 Hz
Coolant:	Oil	Process time:	t = 144 s; t_{ges} = 432 s
Infeed force:	F = 360 N	Belt feed:	v_{fb} = 25.6,...,51.2 mm/min

Figure 10. Experimental setup – Microfinishing of flat surfaces. (a) Kinematics. (b) Surface roughness depending on the used grain size in subsequent process steps. (c) Mirror-like quality of the finished flat surfaces.

First results confirm the possibility to reach a smooth surface by replacing the rotational motion with a tangential motion. Similar to the experiments conducted with rotationally symmetrical specimens, the surface roughness of the flat specimens can be decreased to a maximum height of profile Rz < 0.5 µm by applying three subsequent process steps with decreasing grain sizes. Similar to the test with rotationally symmetrical specimens, the surface roughness of the flat specimens can be decreased to a maximum height of profile Rz < 0.5 µm by applying three subsequent process steps with decreasing grain sizes.

Based on the lower tangential velocity of v_{ft} = 5 m/min, it is necessary to increase the process time in order to reach the intended surface roughness. When finishing flat surfaces, the waviness and unevenness of the surface builds a disturbance variable influencing the generated surface quality. Due to the areal contact, the compensation of these deviations is significantly more time-consuming than when finishing rotational symmetrical specimens. Although, first results on finishing flat surfaces have shown great potential. A grinding process conducted prior to the final finishing prevents excessive process times for the microfinishing process. This combination of grinding and microfinishing is one of the topics for further necessary research.

3.2. Grinding of Geometric Regular and Free-Formed Surfaces

In this chapter, the grinding of surfaces with an increased degree of complexity is analyzed. Possible applications of free-formed coated surfaces are e.g., dies and molds. In first consequence, the workpiece spectrum was enhanced by extracting two dimensional surface contours to increase the degree of complexity, as shown in Figure 11a. As applied for flat surfaces, cylindrical mounted points can be applied for this process. However, the grinding process is limited by the cantilever length of the tool and the radii of the contour. The length of the tool is limited by the small shaft diameter. A long cantilever length leads to a lower stiffness of the tool and a higher tool deflection, resulting in surface defects and shape inaccuracies. Hence, in the performed investigations surface contours extracted by 30 mm were successfully ground. Additionally, the roughness Rz remained constant along the surface contour independent from the varying engagement situations during the grinding of the convex and concave parts of the surface.

Cylindrical mounted points can also be applied for the grinding of rotation-symmetric workpieces, e.g., coated dies for cupping processes (Figure 11b). Here, the surface was ground with a helix tool path along the workpiece cross-section. Based on the different contact areas between grinding tool and workpiece, the machining process was separated into three parts. For the grinding of the flat surface at the top and the cylindrical surface in the middle of the die, a width of cut $a_p = 2$ mm can be applied. However, the width of cut has to be reduced to 0.2 mm for machining the radius section, because of the faceting of the ground surface.

In order to grind workpieces with a larger surface, the use of conical mounted points is one possible solution. For example, the grinding of coated tubes, as seen in Figure 11c, was analyzed during the investigations carried out at the machining center. The grinding tool has to be dressed with the same angle, as the tool is oriented to the workpiece axis of rotation. Applying this process strategy, the process parameter values can be chosen according to the grinding of flat surfaces with cylindrical mounted points. As a consequence, a constant roughness values of Rz = 6.2 μm was measured along the generated surface.

Figure 11. Grinding of (a) regular, (b) rotation-symmetric, and (c) cylindrical geometries.

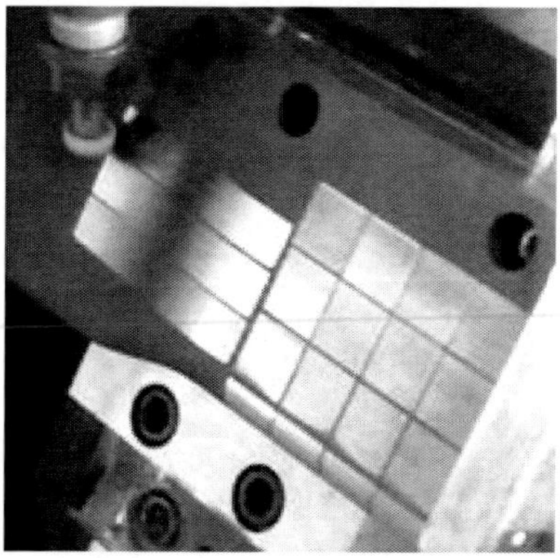

Figure 12. Grinding of free-formed geometries.

The roughness values were measured orthogonal to thecutting direction by the dressing process. As common for milling processes, the tool paths are calculated by conventional computer aided manufacturing (CAM) software tools, like e.g., *HyperMill* or *Cimatron*. In contrast to conventional surface grinding, the width of cut $a_p \leq 0.2$ mm is very small due to the contact area between grinding tool and workpiece surface.

By applying this process strategy, nearly every surface shape can be ground by the use of spherical and toroid shaped mounted points. The surface roughness can be adjusted by the grain size of the tool on the one hand, and by the distance of the width of cut on the other hand.

CONCLUSION

In order to reach the continuously increasing demands placed on tribologically stressed slide faces, new materials as well as an adaption of the abrasive processes used to machine these materials have to be developed. In this paper, aspects and possibilities for the machining of tungsten carbide coatings are shown. A grinding process with special mounted points is able to machine flat, cylindrical, and free-formed surfaces. By using adjusted process parameter values, surface qualities with a maximum height of profile $Rz = 3.4$ μm can be generated. As an alternative, the microfinishing process offers higher surface qualities with a limitation to cylindrical and flat geometries. Regarding the surface topography, very smooth surfaces or plateau-like bearing surfaces can be achieved. When using further process steps with subsequently smaller grain size, it is possible to generate mirror-like surfaces with a maximum height of profile $Rz < 0.3$ μm.

ACKNOWLEDGMENTS

The scientific results presented in this paper concerning the grinding on machining centers are based on investigations of the research project A5 *"Simulation supported NC-shape grinding as a finishing operation of thermally coated deep drawing tools"*, which is kindly funded by the German Research Foundation (DFG) within the Collaborative Research Center (SFB) 708 *"3D-Surface Engineering of Tools for Sheet Metal Forming Manufacturing, Modeling, Machining"*. The investigations and scientific results concerning the microfinishing process described in this paper are based on the research project *"Increasing Power Density and Durability of Highly Stressed Slide Faces by an Improved Surface Integrity Generated with Adapted Machining Processes"*, which is funded by the Mercator Research Center Ruhr Initiative. The authors also would like to thank the Chair of Material Engineering at TU Dortmund University, Germany, for preparing the thermally sprayed coatings.

REFERENCES

[1] Tillmann, W., Vogli, E., Baumann, I., Krebs, B., Nebel, J. (2010). Wear-protective cermet coatings for forming tools. *Materialwissenschaft und Werkstofftechnik,* 41/7, 597-607.

[2] Barletta, M., Bolelli, G., Bonferroni, B., Lusvarghi, L. (2010). Wear and corrosion behavior of HVOF-sprayed WC-CoCr coatings on Al alloys. *Journal of Thermal Spray Technology*, 19/1-2, 358-367.

[3] Chivavibul, P., Watanabe, M., Kuroda, S., Kawakita, J., Komatsu, M., Sato, K., Kitamura, J. (2008). Development of WC-Co Coatings Deposited by Warm Spray Process. *Journal of Thermal Spray Technology*, 17/5-6, 750-756.

[4] Murthy, J. K. N., Rao, D. S., Venkataraman, B. (2001). Effect of grinding on the erosion behaviour of a WC–Co–Cr coating deposited by HVOF and detonation gun spray processes. *Wear*, 249/7, 592-600.

[5] Uhlmann, E., Klein, T., Hochschild, L., Bäcker, C. (2011). Influence of Structuring by Abrasive Machining on the Tribological Properties of Workpiece Surfaces. Procedia Engineering, 19, 363-370.

[6] Jeng, Y. R. (1996). Impact of plateaued surfaces on tribological performance. *Tribology Transactions*, 39/2, 354-361.

[7] Bonny, K., Baets, P., Quintelier, J., Vleugels, J., Jiang, D., Biest, O., Lauwers, B., Liu, W. (2010). Surface finishing: Impact on tribological characteristics of WC-Co hardmetals. *Tribology International*, 43/1-2, 40-54.

[8] Biermann, D., Mohn, T., Blum, H., Kleemann, H. (2010). Grinding of Arc-Sprayed Tungsten Carbide Coatings on Machining Centers: Process Configuration and Simulation. *Key Engineering Materials*, 438, 115-122.

[9] Rausch, S., Biermann, D. (2012). Grinding of hard-material-coated forming tools on machining centers. *Procedia CIRP*, 1, 388-392.

[10] Brinksmeier, E., Giwerzew, A. (2003). Chip formation mechanisms in grinding at low speeds. *Annals of the CIRP*, 52/1, 153-258.

[11] Feng, B. F., Zheng, L., Wang, P., Gai, Q. F., Li, D., Cai, G. Q. (2008). Grinding at very low speed. *Advanced Materials Research*, 53-54, 231-235.

[12] Brinksmeier, E., Heinzel, C., Meyer, L. (2001). Coolant supply conditions and their effect on the workpiece surface layer in grinding. Production Engineering. *Research and Development*, 8/1, 9-12.

[13] Deng, Z. H., Zhang, B., Cheng, F. (2006). Investigations of grinding forces for nanostructured WC/12Co coatings. *Key Engineering Materials,* 304-305, 151-155.

[14] Linke, B. (2008). Dressing process model for vitrified bonded grinding wheels. *Annals of the CIRP*, 57/1, 345-348.

[15] Tawakoli, T., Rasifard, A., Vesali, A. (2009). Effect of the coolant lubricant type and dress parameters on CBN grinding wheels performance. *Advanced Materials Research*, 76-78, 163-168.

[16] De Chiffre, L., Lonardo, P., Trumpold, H., Lucca, D. A., Goch, G., Brown, C. A., Raja, J., Hansen, H. N. (2000). Quantitative Characterisation of Surface Texture. *Annals of the CIRP*, 49/2, 635-652.

[17] Varghese, B., Malkin, S. (1998). Experimental Investigation of Methods to Enhance Stock Removal for Superfinishing. *Annals of the CIRP*, 47/1, 232-234.

[18] Chang, S. H., Balasubramhanya, S., Chandrasekar, S., Farris, T. N., Hashimoto, F. (1997). Forces and specific energy in superfinishing of hardened steel. *Annals of the CIRP*, 46/1, 257-260.

[19] Westkämper, E., Michel, S. (1995). Film finishing Technological investigation and process monitoring. *Production Engineering*, 2/2, 11-14

[20] Puthanangady, T. K. (1995). Experimental investigation of the superfinishing process. *Wear*, 185/1-2, 173-182.

[21] Khellouki, A., Rech, J., Zahouani, H. (2007). The effect of abrasive grain's wear and contact conditions on surface texture in belt finishing. *Wear*, 263/1-6, 81-87.

[22] Khellouki, A., Rech, J., Zahouani, H. (2010). The effect of lubrication conditions on belt finishing. *International Journal of Machine Tools and Manufacture*, 50/10, 917-921.

[23] Rech, J., Kermouche, G., Grzesik, W., García-Rosales, C., Khellouki, A., García-Navas V. (2008). Characterization and modelling of the residual stresses induced by belt finishing on a AISI52100 hardened steel. *Journal of Materials Processing Technology*, 208/1-3, 187-195.

[24] Sacks, J., Welch, W. J., Mitchell, T. J., Wynn, H. P. (1989). Design and Analysis of Computer Experiments. *Statistical science* 4 (4), 409-423.

[25] Khellouki, A., Rech, J., Zahouani, H. (2007). Influence of the belt-finishing process on the surface texture obtained by hard turning. *Journal of Engineering Manufacture*, 221/7, 1129-1137.

[26] De Chiffre, L., Kunzmann, H., Peggs, G. N., Lucca, D. A. (2003). Surfaces in Precision Engineering, Microengineering and Nanotechnology. *Annals of the CIRP*, 52/2, 561-577.

[27] Bruzzone, A. A. G., Costa, H. L., Lonardo, P. M., Lucca, D. A. (2008). Advances in engineered surfaces for functional performance. *Annals of the CIRP*, 57/2, 750-769.

[28] Goeke, S., Brüggemann, T., Biermann, D. (2013). Development of Machining Processes for the Use of Multilayer High-Performance Coatings. *ISRN Tribology*, 1, 1-6.

[29] Zhong, Z. (2001). Machining of thermally sprayed WC-Co coatings. *Materials and Manufacturing Processes*, 16/1, 103-112.

In: Recent Advances in Abrasives Research
Editor: Dirk Bähre
ISBN: 978-1-62808-566-2

Chapter 9

INFLUENCES ON THE PRECISION HONING OF SMALL BORES FOR UNHARDENED STEEL

*Christina Schmitt**
Saarland University, Institute of Production Engineering, Saarbruecken, Germany

ABSTRACT

Honing is an abrasive machining process used to improve the geometric and form accuracy as well as the surface quality of pre-machined workpieces. Applications of honing can be found throughout the metalworking industry and beyond. Examples for the precision honing of small bores are components of fuel injection systems or hydraulic parts. As the honing process is usually one of the last steps in the production process, the process stability and repeatability concerning the product quality are very important. The achievement of a reliable process with honing from the very first part is very difficult because of the many influences to the honing process. Due to high plastification of the material it is especially difficult if unhardened steel is honed. This chapter tries to give an overview of the different influences and shows some results regarding how they influence the process for unhardened workpieces.

Keywords: Honing, friction, force measurement, lubrication, influences

---

* Corresponding author. Email: Christina.schmitt@mx.uni-saarland.de.

1. Introduction

The honing process consists of three overlaying movement components. The honing tool is rotating around its axis and at the same time moving up and down along the axis. The third movement component is the radial feeding movement of the honing stone, which is the actual abrasive part of the tool. Figure 1 shows a typical tool for the honing of small bores (diameter of less than 15mm). In addition to the honing stone, there are two guiding stones. These are to guide the tool in the bore, but are not supposed to participate in the abrasive process. Figure 1 also shows how the feeding movement of the tool works. The axial movement of the feeding cone is translated into a radial movement of the honing stone at the cone surface.

The honing stone itself consists of abrasive grains, such as diamond or cubic boron nitride (CBN), embedded in metallic bonding. Through honing, surface roughnesses R_z less than 1µm can be obtained in serial production. Geometry and form accuracy of less than 1µm can also be obtained. Honing can work with all industrial used material, from metals like hardened or unhardened steels to polymers, glass and ceramics [2].

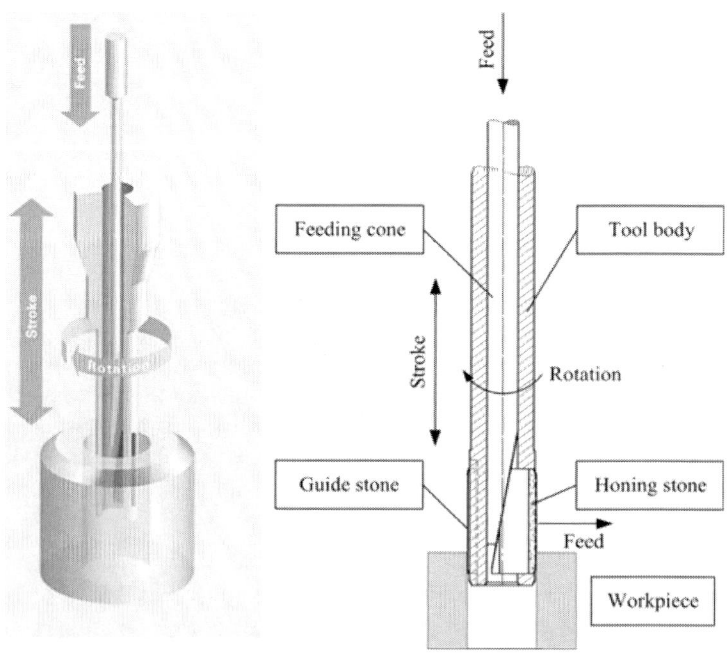

Figure 1. Principle of the honing process [1].

2. INFLUENCES ON THE HONING PROCESS

The honing process has many influences that work on it. Figure 2 gives an overview of these influences. Many of them have been evaluated in past studies, others need some more research. An overview of the state of the art will be given here on different aspects.

The first process influence to be discussed can be seen under the term 'System'. Its most important part is the machine used. One of the main characteristics of a honing machine is the kind of feeding principle it uses. Older machines used hydraulic feeding systems; newer ones have electromechanical systems (Figure 3). The hydraulic systems allow a constant oil pressure and thus a constant cutting pressure from the honing stone on the workpiece. But the difficult bore adjustment and the high maintenance costs have a negative impact. Therefore, the electromechanical systems have gained more and more acceptance. They allow an accurate bore control as the radial position of the honing stone is known at each point in time. In the initial variant, the feed controlled approach, the stone is fed in certain time intervals by certain steps until the final position is reached. But one disadvantage is that the forces change strongly during the process.

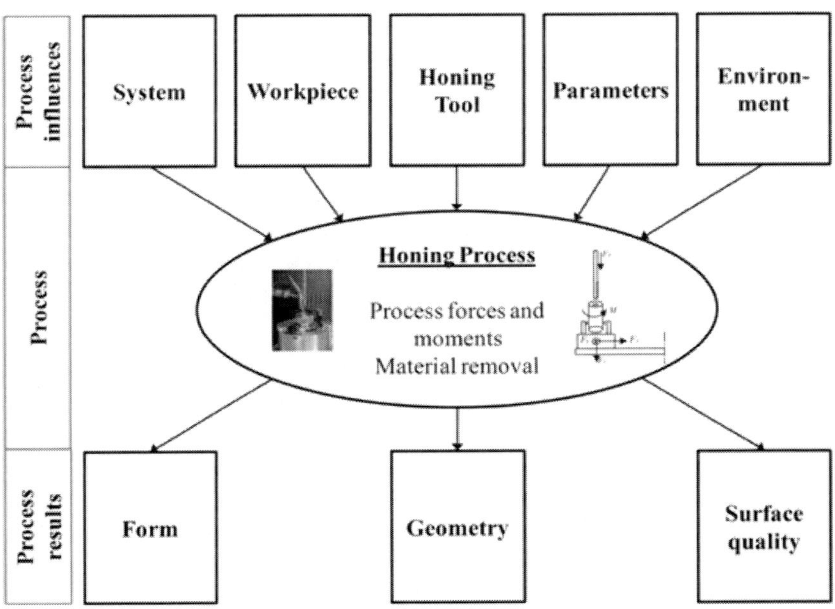

Figure 2. Influences and results of the honing process.

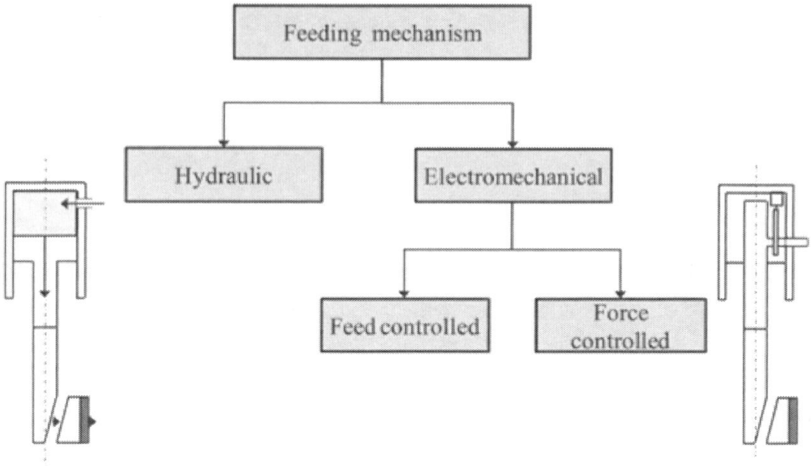

Figure 3. Feeding mechanisms for honing machines.

A more recent approach tries to combine the advantages of hydraulic and electromechanical feed systems [3]. The forces during honing are controlled by a force sensor and kept as constant as possible by an adjusted feeding movement. The feed intervals, -steps and -velocity can vary depending on the control approach [4]. By the more constant force progression, force controlled honing provides the potential to allow better surface qualities and less stresses of workpiece and tool. Figure 4 shows an example of the forces (axial force measured using a force sensor underneath the workpiece and cone force measured using a machine integrated sensor above the feeding cone) for the two control strategies. The upper forces show a more constant course, the variations of the signals are much higher for the feed controlled approach.

For the force controlled approach, the honing machines have to be equipped with a force sensor that allows the control of the process force. For the honing of small bores the latter is mounted above the expanding rod and thus measures only indirectly the requested control size, namely the force from the honing stone on the workpiece wall. Therefore, for a reasonable control the transmission of the forces between feed motor and workpiece has to be known. In [5], an approach for the calculation of the force transmission is shown.

Another important factor for the performance of the honing process is the used cooling lubricant. It should be the right fit for the workpiece material and the honing stone. The tasks of the cooling lubricant are the lubrication of the abrasive material removal process, the cooling of workpiece and tool and the cleaning of the bore from removed material.

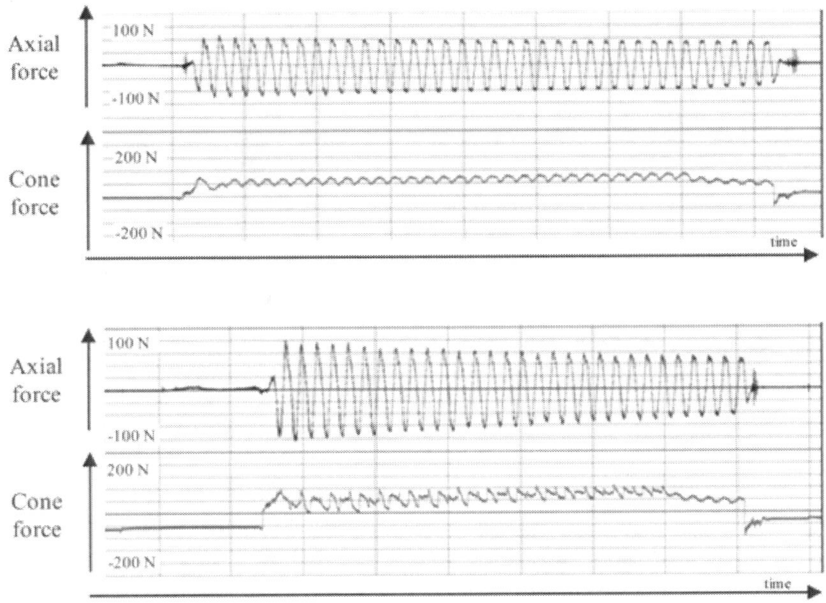

Figure 4. Process forces for force controlled (above) and feed controlled (below) honing.

If the cooling lubricant fulfills these tasks, it can contribute to the precision of the process, improve the cleanness of the workpiece and reduce tool wear. For the honing of unhardened steel it is also an important task of the cooling lubricant to avoid the clogging of the honing stone.

As already stated, honing can work on many materials, ranging from metals like hardened or unhardened steel to other metals like aluminum to nonmetal materials like polymers, glass and ceramics [2]. The workpieces can also have a vast range of dimensions starting from bore diameters of 0.3mm up to several 1000mm. The workpiece material is of course important for the process and influences the choice of the abrasive grains, the cooling lubricant and the process parameters. The quality of the premachining processes is also important for the honing process. If the tolerances are too high, the honing step with its comparatively small amount of material removal is not able to eliminate them.

Figure 5 shows a typical honing tool for the honing of small bores. The tool is equipped with one honing stone and two guiding stones. For bigger bores usually tools with several honing stones are used, for smaller bores galvanically plated tools are used.

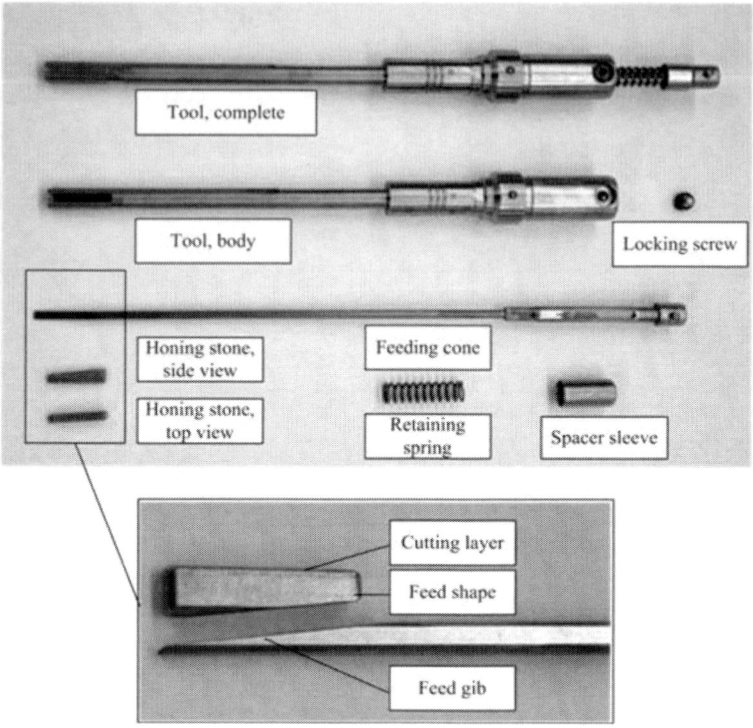

Figure 5. Single-stone honing tool [1].

The most important part of the tool is the honing stone, which is responsible for the material removal. Type and size of the abrasive grains should be adjusted to the work piece material and the desired surface roughness. Also the bonding material, the shape of the grains (e. g. round or sharp-edged) and the concentration of grains play an important role. The influence of the guiding stones with their specification is also important and has to be chosen carefully.

The lifetime of a honing stone can vary from just a few parts up to several thousand machined parts. Honing stones are usually sintered, which leads to a random distribution of grains in the honing stone and to the fact that even two honing stones from the same load can be different because of a different distribution of grains. Up to now there is no comprehensive characterization that can predict the lifetime of a honing stone. During the honing process there is not only material removal of the workpiece, but also of the honing stone. This is important because it enables a self-sharpening effect. Dull grains break out of the bonding and enable new grains to remove material.

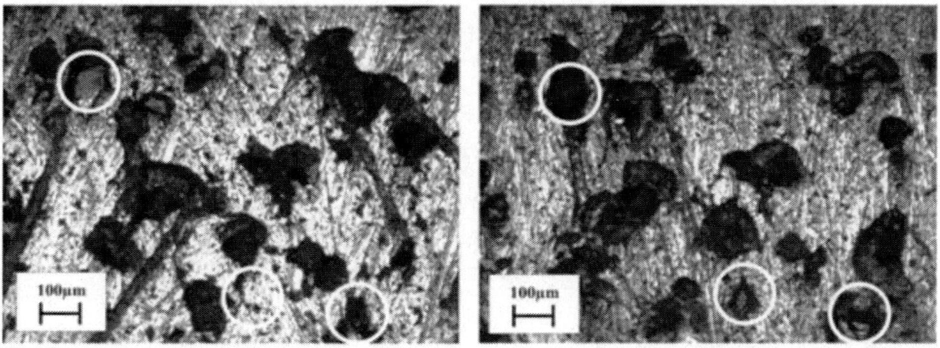

Figure 6. CBN honing stone before (left) and after (right) the honing of 10 parts.

Figure 6 shows the same position on a honing stone equipped with CBN grains in metallic bonding. Between the two pictures 10 parts have been honed. The pictures taken with a laser scanning microscope (LSM) show the change of the honing stone surface.

While the two grains in the upper left corner have apparently fallen out of the bonding and left holes behind, there are two new grains coming to the surface in the lower right corner. The change can be seen over the whole honing stone. The pictures show an area of 1.23mm^2, the whole honing stone has a surface of approximately 60mm^2. These pictures show that even the same honing stone is constantly changing its surface and also its behavior during the honing process.

The honing parameters have a significant influence on the process. The most important parameters are shown in Table 1: The movement parameters rotation speed, lifting speed and lifting acceleration of the tool influence the process strongly. The combination of rotation and lifting movement is responsible for the characteristic crosshatch pattern of a honed surface. The resulting honing angle α between the crosshatch lines is dependent on the relation of the two speed components for rotation and oscillation. The feed per step Z together with the time between feeding steps P characterizes the feeding movement for feed controlled honing, whereas the honing force is the target value for the constant force in force controlled honing. The lifting length has to be adjusted to bore and honing stone length, otherwise a good result regarding the form of the workpiece is impossible. The sparking out time is an optional phase at the end of the honing process, where rotation and lifting movement go on without feed movement. This is supposed to relax the system and thereby gives better results regarding the surface quality of the workpiece. The cone angle φ describes the tool geometry.

Table 1. Honing parameters

Symbol	Unit	Description
v_a	mm/s	Lifting speed
a	mm/s2	Lifting acceleration
n	U/min	Rotation speed
Z	mm	Feed per step
P	s	Time between feeding steps
F_h	N	Honing Force
l_B	mm	Bore length
l_H	mm	Lifting length
l_h	mm	Length of the honing stone
t_A	s	Sparking out time
φ	°	Cone angle

The environment also has an influence on the honing process. The most important factors here are the operator of the machine and its behavior when reacting to process variations, the temperature and vibrations. As the honing process tries to reach tolerances below 1µm, the variation of the temperature in small ranges can have important influence on the results. Therefore, a cooling lubricant is usually used during honing.

3. Investigation of the Honing Process for Unhardened Steel

3.1. Experimental Setup for the Honing of Small Bores

The investigations were performed by using a Kadia vertical honing machine equipped with one spindle for honing, a pneumatic measuring device and a deburring station.

Figure 7 shows the experimental setup with the gimbal-mounted workpiece and the force measuring device underneath it. The honing tool used was equipped with one honing stone and two guiding stones. In the original setup the honing stone with the denomination D15/503/13/50 had diamond grains with an average diameter of 15µm in metallic bonding, the guiding stones also had diamond grains with a smaller average diameter of 10µm in metallic bonding.

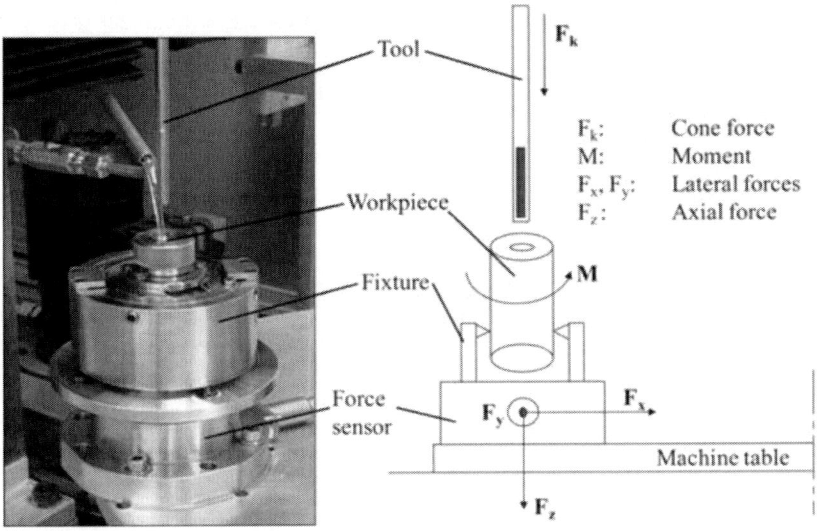

Figure 7. Experimental setup for the honing process [3].

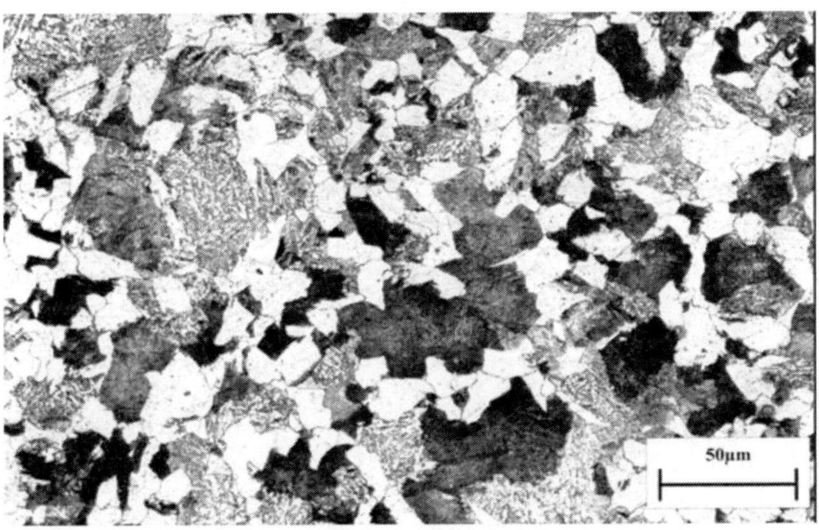

Figure 8. Microstructure of the workpieces made from unhardened steel.

The honed parts were made of unhardened steel (20MnCr5, 1.7147) with an inner diameter of 7.96mm, an outer diameter of 25mm and a length of 50mm. Figure 8 shows the microstructure of the material, which has a hardness of HRC 20. The white phase in the picture is ferrite, the darker areas consist of perlite.

Table 2. Process parameters for the honing experiments

Parameter	Feed controlled	Force controlled
Rotation speed [r/min]	1000	
Lifting speed [m/s]	0.26	
Feed per step [mm]	0.0005	-
Time between fees steps [s]	0.5	-
Honing force [N]	-	80
Honing grains	Diamond	
Honing oil	Kadiol 56	

The process parameters used during the experiments are shown in Table 2. There have been no changes to these if not mentioned otherwise.

3.2. Comparison of Feed and Force Controlled Honing with Regard to Clogging

An important problem while honing soft materials is the so-called clogging effect. This is a process where material from the workpiece clogs on to the surface of the honing stone. Figure 9 shows an example.

The above described honing stone with diamond grains in metallic bonding can be seen as it should be on the right side of the picture. There are grains visible in the metallic bonding, which has metallic gray and bronze color areas because of its ingredients. On the left side, the same honing stone is clogged and by this unusable.

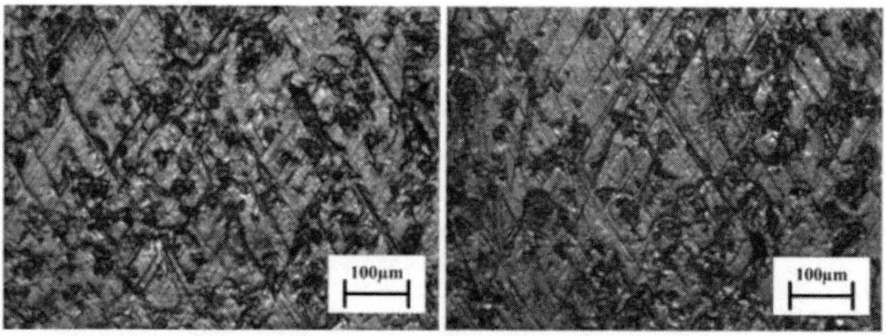

Figure 9. Honing stone with diamond grains for unhardened steel, left: clogged after feed controlled honing, right: free after force controlled honing.

The mechanisms that lead to the clogging effect of the honing stone surface are not fully understood until now. Probably, the strain hardening of the removed workpiece material plays an important role together with the adhesion between tool and workpiece. The conducted experiments point out that the clogging effect involves high honing forces. Figure 9 undermines this. It shows the same honing stone which was first used for feed controlled honing of 25 parts. After this, the honing stone was clogged as shown on the left side of Figure 9, but not yet destroyed. In the next step, 25 parts have been honed force controlled using the same honing stone without any treatment. After this, the honing stone was free again as shown on the right side of Figure 9. The explanation for this behavior is in the lower forces that occurred during force controlled honing. The axial force, measured by the force measuring device underneath the workpiece, was on average 360N during the feed controlled experiments in comparison to 140N during the force controlled ones. A downside to force controlled honing is the longer process time that it usually takes, but the potential against clogging is clearly visible.

3.3. Investigation of the Process Load on Honing- and Guiding Stones

The tool shown above has been used for further experiments. In the end it was destroyed by the clogging effect. Figure 10 shows the damage at the honing- and guiding stones.

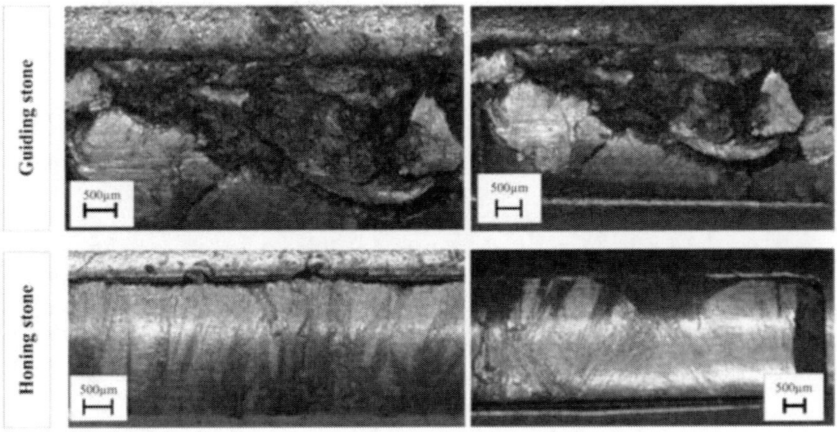

Figure 10. Damage on a honing tool because of the clogging effect.

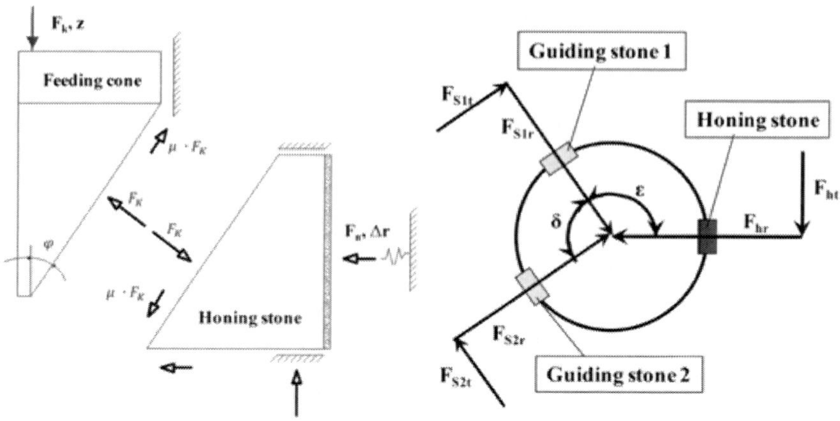

Figure 11. Left: Force transmission between feeding cone and honing stone [6]; right: Forces at the circumference of the honing tool [5].

As can be seen, the cause for the tool damage was the clogging at one of the guiding stones that led to the breaking off of some parts of the guiding stone. These parts of the guiding stone got between tool and workpiece and by this also destroyed the honing stone.

During the honing process the stones have to bear high forces. The forces are different at every stone because of their unbalanced distribution around the circumference of the tool (Figure 11). The forces can be calculated dependent on the force F_k above the feeding cone measured by an internal force sensor in the machine [5].

The radial force at the honing stone can be calculated using this cone force and the tool geometry (ε: angle between honing stone and guiding stone, δ: angle between the two guide stones and cone angle φ between feeding cone and honing stone as shown in Figure 11).

The moment M around the tool axis follows from the tangential forces F_{ht} at the honing stone and F_{s1t} and F_{s2t} at the guiding stones together with the workpiece diameter r_w:

$$M = r_w \cdot (F_{ht} + F_{s1t} + F_{s2t}) \qquad (1)$$

The tangential force components F_{ht}, F_{s1t} and F_{s2t} can be calculated from the radial force components using the tangential force coefficients. They will be further examined in 3.4. With the help of the tangential force ratio, the relation between the normal and tangential forces at the honing stone can be calculated [7, 8]. This gives:

$F_{ht} = \mu_h \cdot F_{hr}$ with $F_{hr} = f(F_k, \text{tool geometry})$ (2)

The same can be supposed for the guiding stones.

$F_{s1t} = \mu_{s1} \cdot F_{s1r}$ and $F_{s2t} = \mu_{s2} \cdot F_{s2r}$ (3)

If two identical guide stones are used, as it is usually done, it can be assumed that $\mu_{s1}=\mu_{s2}=\mu_s$. This gives the moment as:

$$M = r_w \cdot (\mu_h \cdot F_{hr} + \mu_s \cdot F_{s1r} + \mu_s \cdot F_{s2r}) \quad (4)$$

The forces at the two guiding stones F_{s1r} and F_{s2r} can be evaluated using the power balance at the honing tool with the assumption of a gimbal-mounted workpiece (no lateral forces).

$$F_{s1r} = \mu_s \cdot F_{s2r} + F_{hr} \cdot \sin(\varepsilon\text{-}\delta) - \mu_h \cdot F_{hr} \cdot \cos(\varepsilon\text{-}\delta) \quad (5)$$

$$F_{s2r} = \frac{(F_{hr} * ((\mu_h - \mu_s) * \sin(\varepsilon - \delta) + (\mu_s * \mu_h + 1) * \cos(\varepsilon - \delta))}{(\mu_s^2 + 1)} \quad (6)$$

To be able to calculate the forces at the honing- and the two guiding stones, the tangential force ratio for honing- and guiding stones must be known. Their examination will be shown in the following.

3.4. Determination of Tangential Force Coefficients

Figure 12 shows an experimental setup on a turning lathe to determine the tangential force coefficients for a certain combination of honing stone, workpiece material and honing oil. The tangential force coefficient is the proportion of tangential and normal force:

$$\mu = \frac{F_t}{F_n} \quad (7)$$

These coefficients are independent of the used cutting speed [5]. They have been examined for honing- and guiding stones as described above and also for guiding stones made from hard metal. These are supposed to have a smaller tangential force coefficient and therefore to be less susceptible to clogging and tool damage, they will also be used in honing experiments in the following. The cooling lubricant used was Kadiol 56, the workpieces were made from unhardened steel 20MnCr5, with a hardness of HRC 20.

Figure 12. Setup for the determination of tangential force coefficients [5].

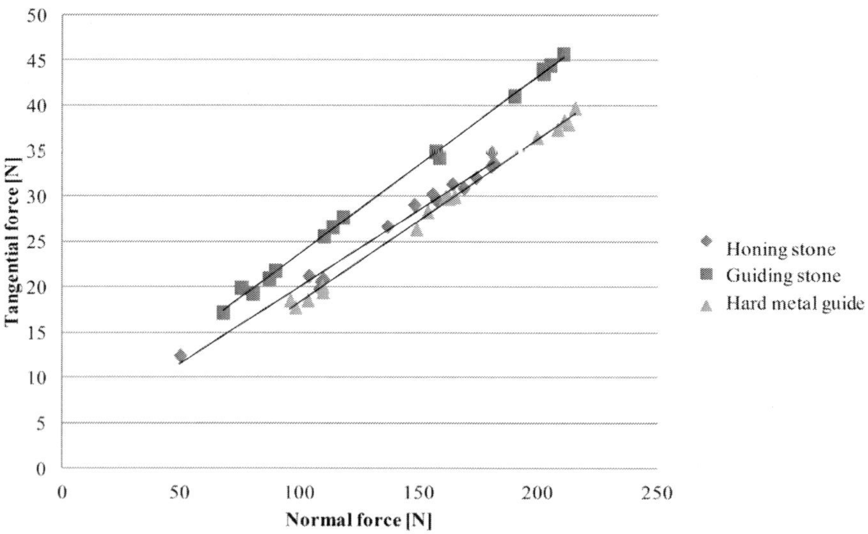

Figure 13. Tangential force coefficients for different honing- and guiding stones.

The results of the experiments in Figure 13 showed that the guiding stones have the highest value of the tangential force coefficient of 0.23, then the honing stone with 0.19 and the hard metal stones with 0.18. The coefficient of determination was above 0.99 for all tangential force coefficients.

3.5. Loads on Different Stones

Using these tangential force ratios, the loads at the honing- and guiding stones can be calculated for certain levels of the cone force F_k. The results are shown in Figure 14 and Table 3. The results show that guiding stone 2 has a much higher load than guiding stone 1 due to its position around the circumference of the tool (Figure 11). During the experiments, guiding stone 2 was also the one which failed and thereby led to the tool damage. As the hard metal guides have theoretically lower load values, they shall be tested in honing experiments in the following.

3.6. Influence of the Lubrication

The influence of different cooling lubricants on the honing process has been investigated in further experiments. The two honing oils tested are Honilo 971 and Kadiol 56. Their characteristics are shown in Table 4.

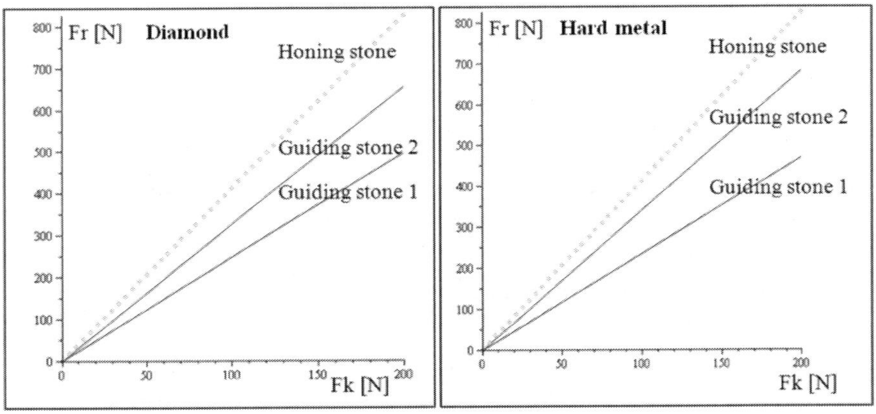

Figure 14. Radial forces on honing- and guiding stones versus the cone force for left: diamond guides, right: hard metal guides.

Table 3. Calculated Loads (radial forces) on honing- and guiding stones

Cone force 100N	Honing stone	Guiding stone 1	Guiding stone 2
Tool with Diamond guides	413N	248N	326N
Tool with hard metal guides	413N	234N	341N

Table 4. Characteristics of the cooling lubricants used

Cooling lubricant	Density at 15°C in kg/cm^3	Kinematic viscosity at 40°C in mm^2/s	Dynamic viscosity in kg/m*s
Kadiol 56	850	5.6	4760
Honilo 971	852	8.4	7157

Table 5. Tangential force coefficients for the different cooling lubricants and stones

Cooling lubricant	Honing stones	Guiding stones diamond	Guiding stones hard metal
Kadiol 56	0.19	0.23	0.18
Honilo 971	0.199	0.188	0.13

Table 6. Calculated normal loads for the different cooling lubricants using guiding stones made from hard metal

Cooling lubricant	Honing stones	Guiding stone 1	Guiding stone 2
Kadiol 56	413N	234N	341N
Honilo 971	413N	216N	357N

The tangential force coefficients were tested for honing stones, guiding stones with diamond grains and guiding stones made from hard metal for both cooling lubricants. The results are shown in Table 5. As guiding stones made from hard metal have lower tangential force coefficients than guiding stones with diamond grains, in the further calculations only hard metal guiding stones are considered.

Table 6 shows the calculated normal forces on the honing stone and the guiding stone for the two different lubricants used. It can be concluded that the higher tangential force coefficients for Kadiol 56 lead to a more balanced force distribution at the guiding stones. Especially the normal load on the higher strained guiding stone 2 is lower for Kadiol 56 in comparison to Honilo 971.

CONCLUSION

This chapter deals with the honing process, a very important process for the finishing of small bores. It is shown that there are many influencing factors to the honing process that have to be known and controlled in order to reach a high process stability and repeatability. This is especially important for the difficult honing of unhardened steels. There, it often comes to the so-called clogging effect which can destroy the honing tools. During clogging the honing- or guiding stone is clogged with workpiece material and in the worst case unusable thereafter. The experiments showed that one of the reasons for the clogging effect can be too high process forces. As shown above, the approach of force controlled honing could be a solution for this problem. As in the conducted experiments the tools all broke due to damage of one of the guiding stones, there has been a closer look at those. The theoretical considerations showed that the guiding stone that gave up was the one with the highest load on it, due to the unequal distribution of honing- and guiding stones around the circumference of the tool. For the calculation, the tangential force coefficients had to be determined with the presented experimental setup. This was done for two different cooling lubricants to test also their influence on the process. Guiding stones made with diamond grains and guiding stones made from hard metal have been tested. A further approach could be the development of better guiding stones, perhaps with a structure on the hard metal guides to lower the forces on the guiding stone and to get a more balanced force distribution. This could help to further improve the honing of unhardened steels.

ACKNOWLEDGMENTS

The author would like to thank the KADIA Produktion GmbH + Co., Nürtingen (Germany) for the provision of their honing machine and for supporting the investigations.

REFERENCES

[1] Schmitt, C., Bähre, D., Forsch, K., Klein, H. (2011). Feinstbearbeiten hochgenauer Bohrungen durch Honen. Jahrbuch Schleifen, Honen, Läppen und Polieren, 65. *Ausgabe* (pp. 259-271), Vulkan-Verlag, Essen.

[2] Klocke, F., König, W. (2005). *Fertigungsverfahren 2 – Schleifen, Honen, Läppen*. Springer Verlag, Berlin.

[3] Bähre, D., Schmitt, C., Moos, U. (2012). Analysis of the differences between Force Control and Feed Control Strategies during the Honing of Bores. *Proceedings of the 5th CIRP Conference on High Performance Cutting*.

[4] Bähre, D., Schmitt, C., Moos, U. (2012). Comparison of different approaches to Force Controlled Precision Honing of Bores. *Proceedings of NAMRI/SME,* Vol. 40.

[5] Schmitt, C., Bähre, D. (2013). An Approach to the Calculation of Process Forces during the Precision Honing of Small Bores. *Proceedings of the 46th CIRP Conference on Manufacturing Systems*.

[6] Mushardt, H. (1986). Modellbetrachtungen und Grundlagen zum Innenrundhonen. *Dissertation*, TU Braunschweig.

[7] Saljé, E., von See, M. (1987). Process-Optimization in Honing. *Annals of the CIRP*, Vol. 36/1, 235-239.

[8] Hoffmeister, H.-W., Menze, B. (2004). Einsatz von Sensorik beim Honen und Regelung von Honprozessen. *Jahrbuch Schleifen, Honen, Läppen und Polieren*, 61. Ausgabe (pp. 261-279), Vulkan Verlag, Essen.

INDEX